Some Physical Constants

Speed of light	c	3.00×10^8 m/s
Gravitational constant	G	6.67×10^{-11} N·m²/kg²
Coulomb's constant	$1/4\pi\varepsilon_0$	8.99×10^9 N·m²/C²
Permittivity constant	ε_0	8.85×10^{-12} C²/(N·m²)
Permeability constant	μ_0	$4\pi \times 10^{-7}$ N/A²
Planck's constant	h	6.63×10^{-34} J·s
Boltzmann's constant	k_B	1.38×10^{-23} J/K
Elementary charge	e	1.602×10^{-19} C
Electron mass	m_e	9.11×10^{-31} kg
Proton mass	m_p	1.673×10^{-27} kg
Neutron mass	m_n	1.675×10^{-27} kg
Avogadro's number	N_A	6.02×10^{23}
Stefan-Boltzmann constant	σ	5.67×10^{-8} W/(m²·K⁴)

Standard Metric Prefixes
(for powers of 10)

Power	Prefix	Symbol
10^{18}	exa	E
10^{15}	peta	P
10^{12}	tera	T
10^{9}	giga	G
10^{6}	mega	M
10^{3}	kilo	k
10^{-2}	centi	c
10^{-3}	milli	m
10^{-6}	micro	μ
10^{-9}	nano	n
10^{-12}	pico	p
10^{-15}	femto	f
10^{-18}	atto	a

Commonly Used Physical Data

Gravitational field strength $g = \lvert\vec{g}\rvert$ (near the earth's surface)	9.80 N/kg $= 9.80$ m/s²
Mass of the earth M_e	5.98×10^{24} kg
Radius of the earth R_e	6380 km (equatorial)
Mass of the sun $M_\odot$	1.99×10^{30} kg
Radius of the sun $R_\odot$	696,000 km
Mass of the moon	7.36×10^{22} kg
Radius of the moon	1740 km
Distance to the moon	3.84×10^8 m
Distance to the sun	1.50×10^{11} m
Density of water[†]	1000 kg/m³ = 1 g/cm³
Density of air[†]	1.2 kg/m³
Absolute zero	0 K $= -273.15$°C $= -459.67$°F
Freezing point of water[‡]	273.15 K = 0°C = 32°F
Boiling point of water[‡]	373.15 K = 100°C = 212°F
Normal atmospheric pressure	101.3 kPa

[†]At normal atmospheric pressure and 20°C.
[‡]At normal atmospheric pressure.

D0224903

Useful Conversion Factors

1 meter = 1 m = 100 cm = 39.4 in = 3.28 ft
1 mile = 1 mi = 1609 m = 1.609 km = 5280 ft
1 inch = 1 in = 2.54 cm
1 light-year = 1 ly = 9.46 Pm = 0.946 × 10¹⁶ m
1 minute = 1 min = 60 s
1 hour = 1 h = 60 min = 3600 s
1 day = 1 d = 24 h = 86.4 ks = 86,400 s
1 year = 1 y = 365.25 d = 31.6 Ms = 3.16 × 10⁷ s
1 newton = 1 N = 1 kg·m/s² = 0.225 lb
1 joule = 1 J = 1 N·m = 1 kg·m²/s² = 0.239 cal
1 watt = 1 W = 1 J/s
1 pascal = 1 Pa = 1 N/m² = 1.45 × 10⁻⁴ psi
1 kelvin (temperature difference) = 1 K = 1°C = 1.8°F
1 radian = 1 rad = 57.3° = 0.1592 rev
1 revolution = 1 rev = 2π rad = 360°
1 cycle = 2π rad
1 hertz = 1 Hz = 1 cycle/s

1 m/s = 2.24 mi/h = 3.28 ft/s
1 mi/h = 1.61 km/h = 0.447 m/s = 1.47 ft/s
1 liter = 1 l = (10 cm)³ = 10⁻³ m³ = 0.0353 ft³
1 ft³ = 1728 in³ = 0.0283 m³
1 gallon = 1 gal = 0.00379 m³ = 3.79 l ≈ 3.8 kg H₂O
Weight of 1-kg object near the earth = 9.8 N = 2.2 lb
1 pound = 1 lb = 4.45 N
1 calorie = energy needed to raise the temperature of 1 g of H₂O by 1 K = 4.186 J
1 horsepower = 1 hp = 746 W
1 pound per square inch = 6895 Pa
1 food calorie = 1 Cal = 1 kcal = 1000 cal = 4186 J
1 electron volt = 1 eV = 1.602 × 10⁻¹⁹ J

$$T = \left(\frac{1\text{K}}{1°\text{C}}\right)(T_{[C]} + 273.15°\text{C}) \qquad T_{[C]} = \left(\frac{5°\text{C}}{9°\text{F}}\right)(T_{[F]} - 32°\text{F})$$

$$T = \left(\frac{5\text{K}}{9°\text{F}}\right)(T_{[F]} + 459.67°\text{F}) \qquad T_{[F]} = 32°\text{F} + \left(\frac{9°\text{F}}{5°\text{C}}\right)T_{[C]}$$

Specific Heats of Various Materials
(at 22°C unless otherwise noted)

Substance	Specific Heat $(J \cdot kg^{-1}K^{-1})$
Elemental Solids:	
Lead	128
Gold	129
Silver	234
Copper	387
Iron	448
Aluminum	900
Other Solids:	
Granite	760
Glass	837
Marble	860
Wood	1700
Ice (−10°C)	2200
Liquids:	
Mercury	139
Ethyl Alcohol	2430
Sea water	3850
Water	4186
Air	≈740

Latent Heats of Various Substances
(at standard pressure)

Substance	Melting Point (K)	Latent Heat of Fusion (kJ/kg)
Hydrogen	14.0	58.6
Oxygen	54.8	13.8
Nitrogen	63.2	25.5
Mercury	234	11.3
Ice	273	333
Lead	601	24.7
Aluminum	933	105
Copper	1356	205

Substance	Boiling Point (K)	Latent Heat of Vaporization (kJ/kg)
Helium	4.2	21
Hydrogen	20.3	452
Nitrogen	77.4	201
Oxygen	90.2	213
Water	373	2256
Mercury	630	296
Lead	2013	853
Aluminum	2720	11400
Copper	2840	4730

Atomic number
Symbol
Atomic mass

1 1A																	18 8A
1 **H** 1.008	2 2A											13 3A	14 4A	15 5A	16 6A	17 7A	2 **He** 4.003
3 **Li** 6.941	4 **Be** 9.012											5 **B** 10.81	6 **C** 12.01	7 **N** 14.01	8 **O** 16.00	9 **F** 19.00	10 **Ne** 20.18
11 **Na** 22.99	12 **Mg** 24.31	3 3B	4 4B	5 5B	6 6B	7 7B	8	9 8B	10	11 1B	12 2B	13 **Al** 26.98	14 **Si** 28.09	15 **P** 30.97	16 **S** 32.07	17 **Cl** 35.45	18 **Ar** 39.95
19 **K** 39.10	20 **Ca** 40.08	21 **Sc** 44.96	22 **Ti** 47.88	23 **V** 50.94	24 **Cr** 52.00	25 **Mn** 54.94	26 **Fe** 55.85	27 **Co** 58.93	28 **Ni** 58.69	29 **Cu** 63.55	30 **Zn** 65.39	31 **Ga** 69.72	32 **Ge** 72.59	33 **As** 74.92	34 **Se** 78.96	35 **Br** 79.90	36 **Kr** 83.80
37 **Rb** 85.47	38 **Sr** 87.62	39 **Y** 88.91	40 **Zr** 91.22	41 **Nb** 92.91	42 **Mo** 95.94	43 **Tc** (98)	44 **Ru** 101.1	45 **Rh** 102.9	46 **Pd** 106.4	47 **Ag** 107.9	48 **Cd** 112.4	49 **In** 114.8	50 **Sn** 118.7	51 **Sb** 121.8	52 **Te** 127.6	53 **I** 126.9	54 **Xe** 131.3
55 **Cs** 132.9	56 **Ba** 137.3	57 **La** 138.9	72 **Hf** 178.5	73 **Ta** 180.9	74 **W** 183.9	75 **Re** 186.2	76 **Os** 190.2	77 **Ir** 192.2	78 **Pt** 195.1	79 **Au** 197.0	80 **Hg** 200.6	81 **Tl** 204.4	82 **Pb** 207.2	83 **Bi** 209.0	84 **Po** (210)	85 **At** (210)	86 **Rn** (222)
87 **Fr** (223)	88 **Ra** (226)	89 **Ac** (227)	104 **Rf** (257)	105 **Db** (260)	106 **Sg** (263)	107 **Bh** (262)	108 **Hs** (265)	109 **Mt** (266)	110	111	112	(113)	114	(115)	116	(117)	

58 **Ce** 140.1	59 **Pr** 140.9	60 **Nd** 144.2	61 **Pm** (147)	62 **Sm** 150.4	63 **Eu** 152.0	64 **Gd** 157.3	65 **Tb** 158.9	66 **Dy** 162.5	67 **Ho** 164.9	68 **Er** 167.3	69 **Tm** 168.9	70 **Yb** 173.0	71 **Lu** 175.0
90 **Th** 232.0	91 **Pa** (231)	92 **U** 238.0	93 **Np** (237)	94 **Pu** (242)	95 **Am** (243)	96 **Cm** (247)	97 **Bk** (247)	98 **Cf** (249)	99 **Es** (254)	100 **Fm** (253)	101 **Md** (256)	102 **No** (254)	103 **Lr** (257)

Six Ideas That Shaped Physics

Unit T: Some Processes Are Irreversible

Third Edition

Thomas A. Moore

McGraw Hill Education

SIX IDEAS THAT SHAPED PHYSICS, UNIT T:
SOME PROCESSES ARE IRREVERSIBLE, THIRD EDITION

This book is printed on acid-free paper.
1 2 3 4 5 6 7 8 9 0 RMN/RMN 1 0 9 8 7 6

ISBN 978-0-07-760096-9
MHID 0-07-760096-7

Senior Vice President, Products & Markets: *Kurt L. Strand*
Vice President, General Manager, Products & Markets: *Marty Lange*
Vice President, Content Design & Delivery: *Kimberly Meriwether David*
Managing Director: *Thomas Timp*
Brand Manager: *Thomas M. Scaife, Ph.D.*
Product Developer: *Jolynn Kilburg*
Marketing Manager: *Nick McFadden*
Director of Development: *Rose Koos*
Digital Product Developer: *Dan Wallace*
Director, Content Design & Delivery: *Linda Avenarius*
Program Manager: *Faye M. Herrig*
Content Project Managers: *Melissa M. Leick, Tammy Juran, Sandy Schnee*
Design: *Studio Montage, Inc.*
Content Licensing Specialists: *Deanna Dausener*
Cover Image: *NASA*
Compositor: *SPi Global*

Dedication

To Allison
whose warmth is legendary

Library of Congress Cataloging-in-Publication Data
Names: Moore, Thomas A. (Thomas Andrew), author.
Title: Six ideas that shaped physics. Unit T, Some processes are irreversible
    /Thomas A. Moore.
Other titles: Some processes are irreversible
Description: Third edition. | New York, NY : McGraw-Hill Education, [2016] |
    2017 | Includes index.
Identifiers: LCCN 2015048202| ISBN 9780077600969 (alk. paper) | ISBN
    0077600967 (alk. paper)
Subjects: LCSH: Irreversible processes—Textbooks.
Classification: LCC QC318.17 M66 2016 | DDC 530—dc23 LC record available at
http://lccn.loc.gov/2015048202

www.mhhe.com

Contents: Unit T
Some Processes Are Irreversible

About the Author

Thomas A. Moore graduated from Carleton College (magna cum laude with Distinction in Physics) in 1976. He won a Danforth Fellowship that year that supported his graduate education at Yale University, where he earned a Ph.D. in 1981. He taught at Carleton College and Luther College before taking his current position at Pomona College in 1987, where he won a Wig Award for Distinguished Teaching in 1991. He served as an active member of the steering committee for the national Introductory University Physics Project (IUPP) from 1987 through 1995. This textbook grew out of a model curriculum that he developed for that project in 1989, which was one of only four selected for further development and testing by IUPP.

He has published a number of articles about astrophysical sources of gravitational waves, detection of gravitational waves, and new approaches to teaching physics, as well as a book on general relativity entitled *A General Relativity Workbook* (University Science Books, 2013). He has also served as a reviewer and as an associate editor for *American Journal of Physics*. He currently lives in Claremont, California, with his wife Joyce, a retired pastor. When he is not teaching, doing research, or writing, he enjoys reading, hiking, calling contradances, and playing Irish traditional fiddle music.

Preface

Introduction

This volume is one of six that together comprise the text materials for *Six Ideas That Shaped Physics*, a unique approach to the two- or three-semester calculus-based introductory physics course. I have designed this curriculum (for which these volumes only serve as the text component) to support an introductory course that combines two elements that rarely appear together: (1) a thoroughly 21st-century perspective on physics (including a great deal of 20th-century physics), and (2) strong support for a student-centered classroom that emphasizes active learning both in and outside of class, even in situations where large-enrollment sections are unavoidable.

This course is based on the premises that innovative metaphors for teaching basic concepts, explicitly instructing students in the processes of constructing physical models, and active learning can help students learn the subject much more effectively. In the course of executing this project, I have completely rethought (from scratch) the presentation of every topic, taking advantage of research into physics education wherever possible. I have done nothing in this text just because "that is the way it has always been done." Moreover, because physics education research has consistently underlined the importance of active learning, I have sought to provide tools for professors (both in the text and online) to make creating a coherent and self-consistent course structure based on a student-centered classroom as easy and practical as possible. All of the materials have been tested, evaluated, and rewritten multiple times. The result is the culmination of more than 25 years of continual testing and revision.

I have not sought to "dumb down" the course to make it more accessible. Rather, my goal has been to help students become *smarter*. I have intentionally set higher-than-usual standards for sophistication in physical thinking, but I have also deployed a wide range of tools and structures that help even average students reach this standard. I don't believe that the mathematical level required by these books is significantly different than that in most university physics texts, but I do ask students to step beyond rote thinking patterns to develop flexible, powerful, conceptual reasoning and model-building skills. My experience and that of other users is that normal students in a wide range of institutional settings can (with appropriate support and practice) meet these standards.

Each of six volumes in the text portion of this course is focused on a single core concept that has been crucial in making physics what it is today. The six volumes and their corresponding ideas are as follows:

Unit C: Conservation laws constrain interactions
Unit N: The laws of physics are universal (**N**ewtonian mechanics)
Unit R: The laws of physics are frame-independent (**R**elativity)
Unit E: **E**lectric and Magnetic Fields are Unified
Unit Q: Particles behave like waves (**Q**uantum physics)
Unit T: Some processes are irreversible (**T**hermal physics)

I have listed the units in the order that I *recommend* they be taught, but I have also constructed units R, E, Q, and T to be sufficiently independent so they can be taught in any order after units C and N. (This is why the units are lettered as opposed to numbered.) There are *six* units (as opposed to five or seven) to make it possible to easily divide the course into two semesters, three quarters, or three semesters. This unit organization therefore not only makes it possible to dole out the text in small, easily-handled pieces and provide a great deal of flexibility in fitting the course to a given schedule, but also carries its own important pedagogical message: *Physics is organized hierarchically*, structured around only a handful of core ideas and metaphors.

Another unusual feature of all of the texts is that they have been designed so that each chapter corresponds to what one might handle in a single 50-minute class session at the *maximum possible pace* (as guided by years of experience). Therefore, while one might design a syllabus that goes at a *slower* rate, one should not try to go through *more* than one chapter per 50-minute session (or three chapters in two 70-minute sessions). A few units provide more chapters than you may have time to cover. The preface to such units will tell you what might be cut.

Finally, let me emphasize again that the text materials are just one part of the comprehensive *Six Ideas* curriculum. On the *Six Ideas* website, at

www.physics.pomona.edu/sixideas/

you will find a wealth of supporting resources. The most important of these is a detailed instructor's manual that provides guidance (based on *Six Ideas* users' experiences over more than two decades) about how to construct a course at your institution that most effectively teaches students physics. This manual does not provide a one-size-fits-all course plan, but rather exposes the important issues and raises the questions that a professor needs to consider in creating an effective *Six Ideas* course at their particular institution. The site also provides software that allows professors to post selected problem solutions online where their students alone can see them and for a time period that they choose. A number of other computer applets provide experiences that support student learning in important ways. You will also find there example lesson plans, class videos, information about the course philosophy, evidence for its success, and many other resources.

There is a preface for students appearing just before the first chapter of each unit that explains some important features of the text and assumptions behind the course. I recommend that *everyone* read it.

Comments about the Current Edition

My general goals for the current edition have been to correct errors, enhance the layout, improve the presentation in many areas, make the book more flexible, and improve the quality and range of the homework problems as well as significantly increase their number. Users of previous editions will note that I have split the old "Synthetic" homework problem category into "Modeling" and "Derivations" categories. "Modeling" problems now more specifically focus on the process of building physical models, making appropriate approximations, and binding together disparate formulas. "Derivation" problems focus more on supporting or extending derivations presented in the text. I thought it valuable to more clearly separate these categories.

The "Basic Skills" category now includes a number of multipart problems specially designed for use in the *classroom* to help students practice with basic issues. The instructor's manual discusses how to use such problems.

I have also been more careful in providing instructors with more choices about what to cover, making it possible for instructors to omit chapters without a loss of continuity. See the unit-specific part of this preface for more details.

Users of previous editions will also note that I have dropped the menu-like chapter location diagrams, as well as the glossaries and symbol lists that appeared at the end of each volume. I could find no evidence that these were actually helpful to students. Units C and N still instruct students very carefully on how to construct problem solutions that involve translating, modeling, solving, and checking, but examples and problem solutions for the remaining units have been written in a more flexible format that includes these elements implicitly but not so rigidly and explicitly. Students are rather guided in this unit to start recognizing these elements in more generally formatted solutions, something that I think is an important skill.

The only general notation change is that now I use $|\vec{v}|$ exclusively and universally for the magnitude of a vector $\vec{v}$. I still think it is very important to have notation that clearly distinguishes vector magnitudes from other scalars, but the old $\text{mag}(\vec{v})$ notation is too cumbersome to use exclusively, and mixing it with using just the simple letter has proved confusing. Unit C contains some specific instruction about the notation commonly used in texts by other authors (as well as discussing its problems).

Finally, at the request of *many* students, I now include short answers to selected homework problems at the end of each unit. This will make students happier without (I think) significantly impinging on professors' freedom.

Specific Comments about Unit T

In the suggested *Six Ideas* unit sequence, this unit is the last of the six. I did this partly so that the course ends with something relatively interesting and easy and partly so that students have seen the quantum physics in unit Q, particularly the material regarding the simple harmonic oscillator and the particle in the box. However, this is not crucial. Students seem to readily accept that quantum systems have energy levels, and this is the most crucial idea. The only *real* requirement is that this unit follow unit C, though the math required for this unit (integral calculus) is more consistent with the level of the math found in units E and Q than in the other units.

Some instructors who have not discussed unit Q may want students to know more about the particle-in-a-box model before discussing chapters T6 and T7 (though I am not sure that this is necessary). If so, then I think that a class session summarizing de Broglie waves and the particle-in-a-box model should be sufficient.

I did not initially intend to revise this unit much for the current edition, but making a few tweaks to improve the flow required other tweaks, and pretty quickly, I found myself reorganizing the unit substantially. The material is mostly the same (except for an entirely new chapter on climate change and a new section on the Stefan-Boltzmann law), but in this edition, I present things in a new order that I hope everyone will consider more logical. The reorganization also gives instructors more flexibility about topics that may be included and omitted.

Chapter T1 now includes a review of heat and work as well as a more historically accurate treatment of treatment of temperature. Instead of taking a detour to discuss gases, chapter T2 now directly introduces the Einstein solid, providing a more streamlined treatment of material that was in chapters T4 and T5 in the previous edition. Chapter T3 now defines entropy and temperature, including material that in the previous edition was in chapters T5 and T6. Chapter T4 is now devoted to the Boltzmann factor (which was only a portion of chapter T6 in the previous edition). This allows me to go into the topic in greater depth, taking advantage of the fact that since the previous edition, online tools such as Wolfram Alpha have made certain difficult integrals more accessible to even beginning students. Chapter T4 now includes some material that was in chapter T7 in the previous edition.

The techniques introduced in chapter T4 now enable a new approach to the ideal gas in chapter T5. Once we have found the heat capacity of both monatomic and diatomic gases, the kinetic-theory calculation from chapter T2 in the previous edition derives the ideal gas law in a more complete and satisfactory way.

Chapter T6 now includes the Maxwell-Boltzmann distribution (from chapter T7 of the previous edition) along with new material about the Stefan-Boltzmann distribution of photon energies in blackbody radiation. This chapter is probably a bit more difficult than the other chapters in the unit (partly because a "distribution" is a pretty subtle concept), but the reorganization also makes it possible to omit this chapter entirely without any loss of continuity (though the material in chapter T6 is helpful for chapter T10).

Chapter T7 provides much the material that was in chapter T3 of the previous edition, but in a form that is both more streamlined and more gently paced. With some regret, I have abandoned the kinetic theory approach to the adiabatic gas law: I found that the length and complexity of that derivation negated the modest insights it offered. Instructors who valued this approach will find it now outlined in one of the homework problems, which I think is a better way for students to learn something of this complexity.

Chapter T8 is mostly the same as before, except that the first section provides a more satisfactory discussion of the entropy of gases. The argument here can be a bit mathematical for some students, but the result in equation T8.7 is what really matters, and one can choose to focus on that. Chapter T9 has also only been lightly revised.

Chapter T10 offers an entirely new exploration of the physics of climate change. I thought that this material provided a strong ending to both the unit and the series, partly because it displays a final rich example of the process of model-building but also because it illustrates how the principles of physics can provide genuine insight into a very important real-world problem. When I teach this chapter, I try to schedule it as the reading assignment for the penultimate class session, so that we can spend the last class session in an open discussion of its implications. This can provide a very powerful and motivating closing experience for students.

I believe that the reorganization makes the whole unit flow much more linearly and logically than before. It also provides instructors with much greater flexibility in deciding what to include and/or omit. Chapters T1 through T4 now comprise the irreducible core of the unit, and by themselves completely address the unit's core question. If you are really short on time, these chapters alone could provide a useful introduction to statistical physics. Adding chapter T5 provides a valuable introduction to ideal gases. Chapter T6 is helpful for chapter T10, but also may be omitted without loss of continuity. I think that (in addition to the core) chapter

T5 is essential for chapter T7, which in turn is essential for chapter T8, which in turn is essential to chapter T9, but one could end the unit after any one of these chapters. Chapter T10 benefits from chapters T1 and T6, but is otherwise completely independent and might be assigned any time after unit C.

Appreciation

A project of this magnitude cannot be accomplished alone. A list including everyone who has offered important and greatly appreciated help with this project over the past 25 years would be much too long (and such lists appear in the previous editions), so here I will focus for the most part on people who have helped me with this particular edition. First, I would like to thank Tom Bernatowicz and his colleagues at Washington University (particularly Marty Israel and Mairin Hynes) who hosted me for a visit to Washingtion University where we discussed this edition in detail. Many of my decisions about what was most important in this edition grew out of that visit. Bruce Sherwood and Ruth Chabay always have good ideas to share, and I appreciate their generosity and wisdom. Benjamin Brown and his colleagues at Marquette University have offered some great suggestions as well, and have been working hard on the important task of adapting some *Six Ideas* problems for computer grading.

I'd like to thank Michael Lange at McGraw-Hill for having faith in the *Six Ideas* project and starting the push for this edition, and Thomas Scaife for continuing that push. Eve Lipton and Jolynn Kilburg have been superb at guiding the project at the detail level. Many others at McGraw-Hill, including Melissa Leick, Ramya Thirumavalavan, Kala Ramachandran, and David Tietz, and Deanna Dausener, were instrumental in proofreading and producing the printed text. I'd also like to thank Dwight Whitaker and Rob Simsiman for reviewing this unit, and students in Physics 70 at Pomona College (especially Juan Zamudio) for helping me track down errors in the manuscript. David Haley helped me with several important photographs. Finally a very special thanks to my wife Joyce, who sacrificed and supported me and loved me during this long and demanding project. Heartfelt thanks to all!

Thomas A. Moore
Claremont, California

SMARTBOOK®

SmartBook is the first and only adaptive reading experience designed to change the way students read and learn. It creates a personalized reading experience by highlighting the most impactful concepts a student needs to learn at that moment in time. As a student engages with SmartBook, the reading experience continuously adapts by highlighting content based on what the student knows and doesn't know. This ensures that the focus is on the content he or she needs to learn, while simultaneously promoting long-term retention of material. Use SmartBook's real-time reports to quickly identify the concepts that require more attention from individual students–or the entire class. The end result? Students are more engaged with course content, can better prioritize their time, and come to class ready to participate.

connect

Learn Without Limits

Continually evolving, McGraw-Hill Connect® has been redesigned to provide the only true adaptive learning experience delivered within a simple and easy-to-navigate environment, placing students at the very center.

- Performance Analytics – Now available for both instructors and students, easy-to-decipher data illuminates course performance. Students always know how they're doing in class, while instructors can view student and section performance at-a-glance.
- Mobile – Available on tablets, students can now access assignments, quizzes, and results on-the-go, while instructors can assess student and section performance anytime, anywhere.
- Personalized Learning – Squeezing the most out of study time, the adaptive engine in Connect creates a highly personalized learning path for each student by identifying areas of weakness, and surfacing learning resources to assist in the moment of need. This seamless integration of reading, practice, and assessment, ensures that the focus is on the most important content for that individual student at that specific time, while promoting long-term retention of the material.

Introduction for Students

Introduction

Welcome to *Six Ideas That Shaped Physics!* This text has a number of features that may be different from science texts you may have encountered previously. This section describes those features and how to use them effectively.

Why Is This Text Different?

Why *active learning* is crucial

Research into physics education consistently shows that people learn physics most effectively through *activities* where they practice applying physical reasoning and model-building skills in realistic situations. This is because physics is not a body of facts to absorb, but rather a set of thinking skills acquired through practice. You cannot learn such skills by listening to factual lectures any more than you can learn to play the piano by listening to concerts!

This text, therefore, has been designed to support *active learning* both inside and outside the classroom. It does this by providing (1) resources for various kinds of learning activities, (2) features that encourage active reading, and (3) features that make it as easy as possible to use the text (as opposed to lectures) as the primary source of information, so that you can spend class time doing activities that will actually help you learn.

The Text as Primary Source

Features that help you use the text as the primary source of information

To serve the last goal, I have adopted a conversational style that I hope you will find easy to read, and have tried to be concise without being too terse.

Certain text features help you keep track of the big picture. One of the key aspects of physics is that the concepts are organized *hierarchically*: some are more fundamental than others. This text is organized into six units, each of which explores the implications of a single deep idea that has shaped physics. Each unit's front cover states this **core idea** as part of the unit's title.

A two-page **chapter overview** provides a compact summary of that chapter's contents to give you the big picture before you get into the details and later when you review. **Sidebars** in the margins help clarify the purpose of sections of the main text at the subpage level and can help you quickly locate items later. I have highlighted technical terms in bold type (like **this**) when they first appear: their definitions usually appear nearby.

A physics **formula** consists of both a mathematical equation and a *conceptual frame* that gives the equation physical meaning. The most important formulas in this book (typically, those that might be relevant outside the current chapter) appear in **formula boxes**, which state the equation, its *purpose* (which describes the formula's meaning), a description of any *limitations* on the formula's applicability, and (optionally) some other useful *notes*. Treat everything in a box as a unit to be remembered and used together.

Active Reading

What is *active reading*?

Just as passively listening to a lecture does not help you really learn what you need to know about physics, you will not learn what you need by simply scanning your eyes over the page. **Active reading** is a crucial study skill for all kinds of technical literature. An active reader stops to pose internal questions such as these: Does this make sense? Is this consistent with my experience? Do I see how I might be able to use this idea? This text provides two important tools to make this process easier.

Features that support developing the habit of active reading

Use the **wide margins** to (1) record *questions* that arise as you read (so you can be sure to get them answered) and the *answers* you eventually receive, (2) flag important passages, (3) fill in missing mathematical steps, and (4) record insights. Writing in the margins will help keep you actively engaged as you read and supplement the sidebars when you review.

Each chapter contains three or four **in-text exercises**, which prompt you to develop the habit of *thinking* as you read (and also give you a break!). These exercises sometimes prompt you to fill in a crucial mathematical detail but often test whether you can *apply* what you are reading to realistic situations. When you encounter such an exercise, stop and try to work it out. When you are done (or after about 5 minutes or so), look at the answers at the end of the chapter for some immediate feedback. Doing these exercises is one of the more important things you can do to become an active reader.

SmartBook (TM) further supports active reading by continuously measuring what a student knows and presenting questions to help keep students engaged while acquiring new knowledge and reinforcing prior learning.

Class Activities and Homework

This book's *entire purpose* is to give you the background you need to do the kinds of *practice* activities (both in class and as homework) that you need to genuinely learn the material. *It is therefore ESSENTIAL that you read every assignment BEFORE you come to class.* This is *crucial* in a course based on this text (and probably more so than in previous science classes you have taken).

Read the text BEFORE class!

The homework problems at the end of each chapter provide for different kinds of practice experiences. **Two-minute problems** are short conceptual problems that provide practice in extracting the implications of what you have read. **Basic Skills** problems offer practice in straightforward applications of important formulas. Both can serve as the basis for classroom activities: the letters on the book's back cover help you communicate the answer to a two-minute problem to your professor (simply point to the letter!). **Modeling** problems give you practice in constructing coherent mental models of physical situations, and usually require combining several formulas to get an answer. **Derivation** problems give you practice in mathematically extracting useful consequences of formulas. **Rich-context** problems are like modeling problems, but with elements that make them more like realistic questions that you might actually encounter in life or work. They are especially suitable for collaborative work. **Advanced** problems challenge advanced students with questions that involve more subtle reasoning and/or difficult math.

Types of practice activities provided in the text

Note that this text contains perhaps fewer examples than you would like. This is because the goal is to teach you to *flexibly reason from basic principles*, not slavishly copy examples. You may find this hard at first, but real life does not present its puzzles neatly wrapped up as textbook examples. With practice, you will find your power to deal successfully with realistic, practical problems will grow until you yourself are astonished at how what had seemed impossible is now easy. *But it does take practice*, so work hard and be hopeful!

Temperature

Chapter Overview

Section T1.1: Introduction to the Unit

Thermodynamics is the study of how a macroscopic object's temperature, thermal energy, entropy, and similar *macroscopic* properties are affected by its interactions with other objects and its general environment. **Statistical mechanics** is the theory that explains the thermodynamic behavior of complex systems in terms of the *statistical* behavior of molecules obeying simple laws of mechanics. We will explore the foundations of statistical mechanics in this unit.

Section T1.2: Irreversible Processes

The simple interactions between molecules at the microscopic level are **reversible processes,** in the sense that a reversed movie of the interaction also depicts a physically possible interaction. Many processes involving macroscopic objects, however, are **irreversible processes** in that sense. One of the challenges faced by statistical mechanics is to explain how irreversible macroscopic behavior arises out of reversible microscopic behavior.

This unit is divided into three major subunits. After this introductory chapter (which lays out the basic questions we will answer), the first subunit uses a simple quantum model of a solid to explore Boltzmann's solution to the problem of irreversibility and to develop useful tools. The second applies these tools to the tougher problem of gases. The third explores important practical consequences of these ideas.

Section T1.3: Heat, Work, and Internal Energy

The technical terms **heat, work,** and **energy transfer** describe energy flowing across a system boundary during a process. The *heat Q* is the part of the energy flow that is driven by a temperature difference between the system and its surroundings; *work W* refers to energy flowing due to external forces acting on the system, and energy transfers [*E*] refer to other kinds of energy flows across the boundary. The energy stored *inside* a system boundary is the system's **internal energy** U. For a system at rest, these quantities are linked by the **first law of thermodynamics:**

$$\Delta U = Q + W + [E] \tag{T1.1}$$

- **Purpose:** This equation expresses the law of conservation of energy in the context of thermodynamic systems, where ΔU is the change in a system's thermal energy in a given process and Q, W, and [E] are the heat, work, and other energy transfers, respectively, that have flowed into or out of the system during that process.
- **Limitations:** Heat, work, and energy transfers are mutually exclusive concepts, but together they include all the ways energy can flow into or out of a system. However, this does assume that the system is at rest.
- **Note:** In this text, both Q and W are positive when energy flows *into* the system. Some other texts adopt a different sign convention for W.

Section T1.4: The Paradigmatic Thermal Process

When a hot object is placed in contact with a cold object, heat flows irreversibly from the former to the latter until they come into **thermal equilibrium** (their thermal properties no longer change with time). Our primary goal in this unit will be to explain this **paradigmatic thermal process.** Understanding this process provides the key to understanding other irreversible processes as well.

Section T1.5: Temperature and Equilibrium

A **thermoscope** is an object having a measurable property whose numerical value is uniquely linked to the object's temperature. Thermoscope measurements empirically show that objects interacting thermally obey the **zeroth law of thermodynamics:**

> A well-defined quantity called **temperature** exists such that two objects in thermal contact will be in thermal equilibrium *if and only if* they both have the same temperature (that is, thermoscope reading).

This expresses the core physical meaning of temperature.

Section T1.6: Thermometers

But how do we define a numerical temperature scale? Scientists first simply selected a thermoscope and defined the temperature at two points (for example, the freezing and boiling points of water) to define the scale. But inconsistencies led Kelvin in 1848 to propose a physical temperature scale that did not depend on any thermoscope (we will see how he did this in chapter T3). Kelvin chose the zero of his scale to be **absolute zero,** the lowest temperature that anything can possibly have, and defined the size of his temperature unit so that 1 kelvin (1 K) = 1°C, or, equivalently, so that the freezing point of water has a temperature of 273.15 K.

A thermometer is a thermoscope calibrated to display temperatures on Kelvin's scale. The pressure of the low-density gas in a **constant-volume gas thermoscope** is almost exactly proportional to Kelvin's scale, as we will see in chapter T5. The section also discusses how to convert between various historical temperature scales.

Section T1.7: Temperature and Thermal Energy

The temperature T of most objects increases as we add energy to an object (increasing its internal energy). We call the part of an object's internal energy that varies with temperature its **thermal energy,** and U will represent this energy in what follows unless otherwise noted. An object's thermal energy U is linked to its temperature by

$$dU = mc\,dT \qquad \text{or} \qquad c \equiv \frac{1}{m}\frac{dU}{dT} \qquad \text{(generally)} \qquad \text{(T1.4)}$$

$$dU \approx \left(\frac{n}{2}k_B\right)N\,dT \qquad \text{or} \qquad \frac{1}{N}\frac{dU}{dT} \approx n\left(\frac{k_B}{2}\right) \quad \left(\begin{array}{c}\text{for certain simple}\\ \text{substances}\end{array}\right) \qquad \text{(T1.5)}$$

- **Purpose:** These equations describe empirically observed links between the change dU in an object's thermal energy and its corresponding change in temperature dT, where m is the object's mass, c is the object's **specific heat** (*not* the speed of light!), N is the number of molecules, n is approximately an integer, and k_B is **Boltzmann's constant** $= 1.38 \times 10^{-23}$ J/K.
- **Limitations:** Both dU and dT must be small enough that $c \approx$ constant, and the object must maintain the same phase, chemical composition and nuclear composition during the process. The second equation applies only to very simple substances (monatomic solids, where $n \approx 6$, low-density monatomic gases, where $n \approx 3$, and diatomic gases, where $n \approx 5$).
- **Notes:** The right-hand version of the top equation basically defines c. Note that *specific heat* is a conventional term, but is a potentially confusing misnomer here, because we can add energy in forms other than heat.

T1.1   Introduction to the Unit

Complex objects have their own distinct physics

In much of this course to date, we have treated objects (from atoms to automobiles to planets) as if they were point-like particles. Yet every object that we experience in our daily lives is in fact a complicated system comprised of an enormous number of tiny particles. We laid some of the foundations for studying complex systems in unit C, but in this unit we will study the behavior of such systems in greater depth.

Complex systems have physical properties that particles do not. We have already seen in unit C how complex systems have *internal energy*, which is related in some (as yet mysterious) way to its *temperature*. In this unit, we will see that complex systems also have something called *entropy*. These are not properties that a simple particle can have.

We will see in each case that such quantities are ultimately *statistical* expressions of the internal "state" of a complex system. For example, the link between temperature and a system's state is vividly illustrated by the freezing of water. Clearly something quite dramatic happens to the internal state of a system of water molecules when we lower its temperature below 0°C.

Thermodynamics and statistical mechanics

The study of how a macroscopic object's temperature, internal energy, entropy, and similar *macroscopic* characteristics are affected by its interactions with other objects and its general environment is called **thermodynamics,** which became a well-developed science in the middle decades of the 1800s, long before physicists understood that macroscopic objects were actually systems of molecules. One of the greatest triumphs of physics in the early decades of the 1900s was the development of **statistical mechanics,** which explains the thermodynamic behavior of systems in terms of the *statistical* behavior of molecules obeying the simple laws of mechanics. We will focus our attention mostly on the principles of statistical mechanics in this unit.

T1.2   Irreversible Processes

Historically, the greatest impediment to the development of statistical mechanics was a particular stark difference between the behavior of macroscopic objects and the behavior of point particles: the behavior of particles is generally **reversible,** while that of macroscopic objects is often **irreversible.**

Reversible processes

Imagine, for example, a video of two perfectly elastic balls colliding on a pool table. The video shows the balls approaching each other, exchanging some energy and momentum during the collision, and then receding from each other. If we run the video backward, we would also see the balls approaching, exchanging some energy and momentum in the collision process, and then departing. As long as the collision is perfectly elastic (so that colliding balls' thermal energies are not involved), we cannot determine from the events depicted whether the video is running forward or backward.

Similarly, imagine a video of someone throwing a ball into the air and then catching it again as it comes down. As the ball goes up, its initial kinetic energy is converted to potential energy. As it comes down, the reverse occurs—its potential energy is converted to kinetic energy. If we run the video of this process backward, we again see a ball going up in the air and coming down: we see nothing strange or unphysical. If a process looks equally plausible in a video run backward as in one run forward, it is **reversible.**

This is not accidental: the laws of physics (either Newtonian or quantum) that describe the interactions of simple particles are completely time-symmetric. If a given simple-particle interaction is physically possible, then

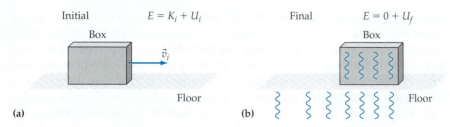

Figure T1.1

The friction interaction between the floor and a sliding box converts kinetic energy to internal energy. This process is *irreversible:* although the reverse process would be consistent with conservation of energy, it is never observed to occur.

a video of that interaction run in reverse also shows a physically possible interaction. These laws of physics imply no forward arrow of time.

Macroscopic objects can behave quite differently. Consider the simple case of a box sliding across a level floor (see figure T1.1). As the box slides, friction slows it down, meaning that its kinetic energy is decreasing. Because the floor is level, the box's *potential* energy does not change in this case. In unit C, we saw that the box's kinetic energy is instead converted to *internal* energy, as indicated by increases in the rubbing surfaces' temperatures.

Irreversible processes

One might think that this is a simple energy transformation process similar to the case of the ball being thrown into the air, but consider a reversed video of this process. In such a video, a box sitting at rest on the floor spontaneously accelerates while the rubbing surfaces get cooler. Viewers unaware that the video is reversed would really sit up in their seats, because we all know that the simple energy transfer process in the reversed video (internal energy to kinetic energy) does *not* occur in nature, even though this conversion would be completely consistent with the conservation of energy. The conversion of kinetic energy to internal energy through friction is **irreversible:** the time-reversed process is *not* physically possible.

My professor in an upper-level thermal physics class once handed each of several groups of us a film movie camera and told us to make a movie illustrating some principles of thermal physics. One group of students (not mine, I regret to say) made their movie and then rewound the film on the spool so that the movie would display in reverse. They cleverly intermixed reversible and irreversible processes so parts of the movie looked completely normal, while other parts were completely outrageous. The movie ended (began?) showing the dust cover of our hardbound thermal physics textbook being miraculously created from a heap of ashes in a fire. One of my favorite sequences began with the image of a placid pond on a beautiful spring day. Strange ripples ominously began to form on the pond's surface. Growing stronger, these ripples converged to a point on the pond, which suddenly disgorged a large rock. The rock leapt out of the pond into the air, falling into the hands of a surprised student. This movie very memorably made the point that some processes are just not physically possible in reverse.

Why is this so? This question becomes even more acute when we seek to explain such behavior in terms of the interactions between microscopic particles, which (as we've seen) are *reversible*. This seems logically absurd! How can reversible microscopic processes lead to irreversible macroscopic ones?

Ludwig Boltzmann was one of the first physicists to really understand that this contradiction was only apparent. During the 1870s, he published a series of fundamental papers that explained the relationship between the energy, temperature, and entropy of macroscopic objects and the microscopic motions of atoms in those objects. His work was energetically criticized by many physicists who doubted the reality of atoms and felt that physical theories should not be based on such unobservable and hypothetical entities. Many also dismissed his work out of hand because they could not see how irreversibility could be consistent with reversible microscopic interactions. As a result of struggling against this criticism, Boltzmann became increasingly despondent, finally committing suicide in 1906 just before experiments with brownian motion and quantization of electric charge made it completely clear that the atomic hypothesis was correct.

This unit's "great idea"

This unit's "great idea" is that Boltzmann was right: *we can explain the irreversible behavior of a complex system by the statistical consideration of the reversible interactions of its molecules.* Our goal in this unit is to develop the ideas and techniques we need to understand this basic idea and its consequences.

The structure of the unit

Figure T1.2 shows how this unit is organized. This first chapter introduces the basic concepts of irreversibility, temperature, and equilibrium. Chapters T2 through T4 represent the unit's conceptual core, using a simple model of a macroscopic object to show how irreversibility arises from reversible microscopic processes, presenting the formal definitions of entropy and temperature, and developing a powerful tool (the Boltzmann factor) that links microscopic to macroscopic physics. Chapters T5 through T7 apply these methods to gases (including photon gases), which adds important new insight. Chapters T8 through T10 then use these methods to explore important practical applications, including heat engines and climate change.

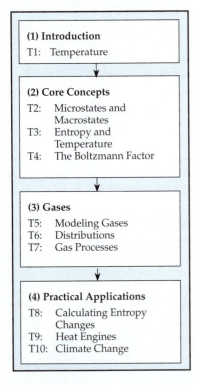

(1) Introduction

T1:   Temperature

(2) Core Concepts

T2:   Microstates and
       Macrostates
T3:   Entropy and
       Temperature
T4:   The Boltzmann Factor

(3) Gases

T5:   Modeling Gases
T6:   Distributions
T7:   Gas Processes

(4) Practical Applications

T8:   Calculating Entropy
       Changes
T9:   Heat Engines
T10:   Climate Change

Figure T1.2

An overview of this unit's structure.

Exercise T1X.1

Which of the following physical processes are (at least approximately) reversible and which are irreversible?
(a) A glass of milk spills on the floor.
(b) An object moves downward, compressing a spring.
(c) Ink poured into a cup of water gradually disperses throughout the water.
(d) A dropped book slams into the floor and remains at rest afterward.
(e) The moon orbits the earth.

T1.3   Heat, Work, and Internal Energy

We introduced the concepts of *heat*, *work*, and *internal energy* in unit C, but these concepts are so important that they are worth reviewing briefly here.

An object's **internal energy** U is the sum of the kinetic energies of all its microscopic particles and the potential energies associated with their interactions. At the macroscopic level, we find that changes in this internal energy are associated with changes in the object's temperature, its phase (whether it is a solid, liquid, or gas), its chemical composition, and/or its nuclear composition. In this unit, we will generally consider only the portion of the internal energy that is linked to changes in an object's temperature: we call this portion of an object's internal energy its **thermal energy.**

In thermal physics, we are often interested less in the total *amount* of an object's total internal energy than in how that energy *changes* under certain conditions. As discussed in unit C, knowing how that energy changes involves watching the amount of energy that crosses the object's boundaries. The crucial thermodynamic concepts of *heat* and *work* both describe *energy transfers across a system boundary.*

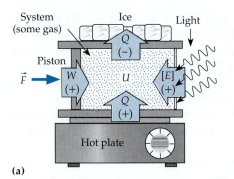

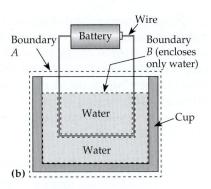

(a)　　　　　　　　　　　　　(b)

Figure T1.3

(a) An illustration of the distinctions between internal energy U, heat Q, work W, and other kinds of energy transfers [E].

(b) How you define a system's boundary makes a difference: energy flows into the system surrounded by boundary A in the form of an electrical energy transfer, but into the system surrounded by boundary B as heat.

When we place a hot object in contact with a cold object, energy spontaneously flows across the boundary between them until both objects come to have the same temperature. At the microscopic level, this transfer takes place because rapidly moving molecules in the hot object transmit some of their energy to the slow-moving molecules in the cold object, either by bumping into them or by some other mechanism. At the macroscopic level, we simply see that the hot object's internal energy decreases and the cold object's internal energy increases. In physics, **heat** is any energy that crosses a boundary between objects *because* of a temperature difference across that boundary. Let me emphasize that to be *heat*, the energy in question *must* (1) be flowing across some kind of boundary between systems (2) as a direct result of a temperature difference across that boundary. In figure T1.3a, the hot plate causes heat to flow *into* the system, increasing its internal energy U, while the cold ice causes heat to flow *out* of the system, decreasing U. In the first case, we say that the heat flow Q is positive; in the second case, Q is negative.

Work is energy that flows across a boundary due to some kind of *external force* acting on the object. If the point on the system where we apply the external force $\vec{F}$ moves a displacement $\Delta \vec{r}$ during a given process, then an energy $W = |\vec{F}||\Delta \vec{r}|\cos\theta$ flows across the system boundary during that process, where θ is the angle between $\vec{F}$ and $\Delta \vec{r}$. In figure T1.3a, if we push on the piston to compress the enclosed gas, we do positive work W on the system, increasing its internal energy. The gas's temperature may increase as it absorbs this energy, but we are not "heating" the gas.

Definition of *work*

We call any *other* kind of energy crossing a system boundary an **energy transfer** [E]. Energy may be carried across a system boundary by electromagnetic waves (for example, light), mechanical waves (for example, sound), electrical currents, or by particles moving across the boundary.

The precise distinction between heat, work, and energy transfers is more a matter of human definition and convenience than one of deep physics: at the microscopic level, a macroscopic energy flow is simply the result of a huge number of microscopic energy transfers between molecules due to collisions or other interactions. Moreover, whether we call a flow of energy in a given process "heat," "work," or an "energy transfer" can depend on our choice of system boundary. For example, consider the situation shown in figure T1.3b. Electricity flows into a piece of resistive wire that is immersed in a cup of water. Electric current flowing in the wire causes its temperature to increase, which in turn causes the water's temperature to increase. If we draw the boundary labeled A (making "the system" the water, the cup, and the resistive wire), then the energy crossing the boundary is an energy transfer carried by an electric current. If we draw the boundary labeled B (making "the system" the water alone), then the energy crossing the boundary is *heat*, driven by the temperature difference between the wire and the water.

The distinctions between Q, W, and [E] can depend on system boundaries

Note, however, that *heat*, *work*, and *energy transfers* do all describe energy in transit *across* a boundary. This sharply distinguishes all three from the object's *internal energy U*, which refers to energy *enclosed by* the system boundary. Because changes in an object's internal energy often affect its temperature, heating an object may make it warmer, but a hot object does not "contain heat"—it simply contains more internal energy.

Any energy flow across the system's boundary can contribute to changes in the object's internal energy. In fact, conservation of energy implies that

The first law of thermodynamics

$$\Delta U = Q + W + [E] \tag{T1.1}$$

- **Purpose:** This equation expresses the law of conservation of energy in the context of thermodynamic systems, where ΔU is the change in a system's thermal energy in a given process and Q, W, and $[E]$ are the heat, work, and other energy transfers, respectively, that have flowed into or out of the system during that process.
- **Limitations:** Heat, work, and energy transfers are mutually exclusive concepts, but together they include all the ways energy can flow into or out of a system. However, this does assume that the system is at rest.
- **Note:** In this text, both Q and W are positive when energy flows *into* the system. Some other texts adopt a different sign convention for W.

We call this crucial equation **the first law of thermodynamics.**

You will become confused by much of what follows if you do not understand the technical distinctions between *heat*, *work*, *energy transfers*, and *internal energy*. It is particularly easy to misuse the word *heat* (even some physics texts do!). The following exercise should help develop your ability to distinguish between these terms.

Exercise T1X.2

In each process described below, energy flows from object A to object B. Is the energy flow involved *heat*, *work*, or another kind of energy transfer? (Write Q, W, or $[E]$ in the blank provided.)

(a) The soup (B) in a pan sitting on an electric stove (A) gets hot.   _____

(b) Light from an incandescent bulb's filament (A) flows to its surroundings (B).   _____

(c) You (A) compress air (B) in a bike pump, making it warm.   _____

(d) Your hands (B) are warmed when they face a fire (A).   _____

(e) The atmosphere (A) warms a reentering spacecraft (A).   _____

(f) A hot pie (A) becomes cooler while sitting in the kitchen (B).   _____

(g) Your chair (B) becomes warmer after you (A) sit in it for a while.   _____

(h) Your cocoa (B) becomes hot in a microwave oven (A).   _____

(i) A drill bit (B) becomes hot after being spun by the drill (A).   _____

T1.4   The Paradigmatic Thermal Process

Many types of processes in nature are irreversible. In the next few chapters, however, we will focus our attention on one particularly simple type of process. Understanding why this particular process is irreversible will provide the key to understanding all other irreversible processes.

Suppose we drop a hot block of metal into a beaker of cold water (see figure T1.4). We can observe (by touching the block and water if necessary) that the block gets cooler and the water gets warmer until their temperatures no longer appear to change. When their temperatures no longer are discernibly changing, we say the two objects are in **thermal equilibrium.** This process is irreversible: one never sees a warm metal block placed in warm water spontaneously get hotter while the water becomes colder.

What is happening here? Colloquially, we might say, "Heat flows from the block to the water." In unit C, we learned what flows here is actually *energy:* as energy flows from the block to the water, the block's thermal energy decreases (making it colder) while the water's thermal energy increases (making it warmer). Since this energy flow is driven by the temperature difference between the objects, it must be *heat.*

This process (a hot object coming into equilibrium with a cold object) will serve for us as the paradigm of an irreversible process. It is simple and commonplace, and yet it raises all the big questions we must address:

1. *What is temperature?* What do we really mean when we say *hot* and *cold?* How is temperature related to thermal energy? How is temperature linked to other properties of an object?
2. *What is heat?* How is it related to temperature and internal energy? Why does heat spontaneously flow from a hot object to a cold object?
3. *What is thermal equilibrium?* Once energy does begin to flow between the objects, why does it stop? How is thermal equilibrium related to temperature?
4. *Why is the paradigmatic process irreversible?* Why does heat spontaneously flow from a hot object to a cold object but never the reverse?

In the next few chapters, we will return repeatedly to this **paradigmatic thermal process** as we address the questions that it raises.

The paradigmatic process

Questions raised by this paradigmatic process

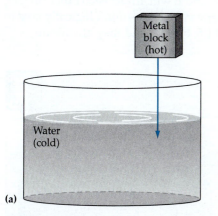 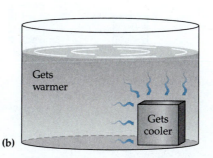 

Figure T1.4

The paradigmatic thermal process. (a) A hot object (a block of metal here) is placed in contact with a cold object (water here). (b) The hot object gets cooler, and the cold object warmer (c) until their temperatures no longer change; at this point, the objects are in *thermal equilibrium.*

T1.5   Temperature and Equilibrium

The first step in making thermal physics a science is to learn how to quantify an object's temperature in a meaningful and reproducible manner. But what *is* temperature, really? In this section and the next, we will create a meaningful quantitative definition of temperature from more primitive ideas.

What we colloquially call "temperature" is something we measure qualitatively by touch: we can feel when an object is *hot* and when it is *cold*, and we can even crudely distinguish degrees of the same. Measurement of temperature by touch is subjective and sometimes inconsistent. Even so, our crude direct sense of temperature allows us to discover that the physical characteristics of many objects change as their temperatures change. For example, as temperature increases, the electric resistance of a conductor changes, the pressure of a confined gas increases, certain kinds of liquid crystals undergo color changes, and almost all substances expand a tiny bit.

Quantifying temperature with a thermoscope

Suppose we have a device that measures a temperature-dependent property of something we can use as a probe and displays a numerical result (for example, an ohmmeter displaying a metal bar's resistance). Imagine that we also verify that (at least over some specified range) the displayed number always increases when we add energy to the probe (say, with a flame) and decreases when we remove energy (say, by putting it in a freezer). We can then call the device a **thermoscope.** A thermoscope is not a *thermometer,* because the value displayed is not *equal* to the probe's temperature: all we know is that the displayed value is *related* to its temperature (maybe in a complicated way) and that unique displayed values correspond to unique temperatures.

The link between temperature and equilibrium

When we bring a thermoscope probe into close physical contact with another object, we notice that the thermoscope display settles down to a fixed number as the probe and object come into equilibrium. Further experiments show that *two objects A and B will be in thermal equilibrium with each other if and only if the value displayed by a thermoscope T in equilibrium with each is the same.* For example, in our paradigmatic process, the values our thermoscope probe displays when we put it in successive contact with the hot metal block and the cold water are initially different but always become equal as the block and water come into equilibrium (see figure T1.5).

This is an empirical result: there is no logical reason that it *must* be true. For example, if an electrically charged object *T* equally attracts charged objects *A* and *B*, it does not follow that *A* and *B* will attract each other (in fact, they will repel one another). Thermoscope values thus don't logically *need* to agree in the situation shown in figure T1.5, but we find they always do.

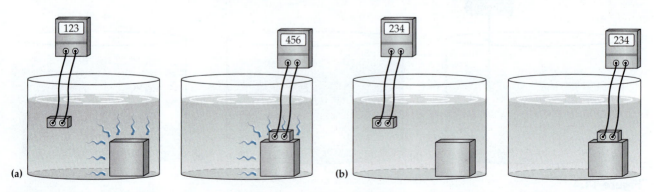

Figure T1.5
Suppose we place a hot metal block into cold water. (a) Before they come into thermal equilibrium, they have different temperatures. (b) When the block and water come into thermal equilibrium, though, both are always found to have the *same* thermoscope reading.

Since values displayed by a good thermoscope are uniquely linked to the quantity that we want to call temperature, we can conclude that:

> A well-defined quantity called **temperature** exists such that two objects in thermal contact will be in thermal equilibrium *if and only if* both have the same temperature.

The zeroth law of thermodynamics

This is called (somewhat whimsically) the **zeroth law of thermodynamics** because physicists now recognize it as being *logically* prior to laws that, when the zeroth law was first discussed, had already been named the "first" and "second" laws of thermodynamics.

This law in fact describes *the* essential physical characteristic of the quantity we call temperature. While the most *apparent* thing about temperature is our sensation of hot and cold, from the perspective of physics, what makes the idea of temperature meaningful and useful is that it *characterizes equilibrium*.

T1.6 Thermometers

As I stated earlier, a thermoscope does not display temperature directly. The values displayed by resistance thermoscopes using different bars of metal, for example, will be entirely different, even under the same conditions. Agreeing on a conventional temperature scale would make communicating scientific results much easier.

But how do we choose a temperature scale? We *could* do this by simply agreeing to select one specific thermoscope to *define* the temperature scale and calibrating all other thermoscopes to agree with it. Historically, this was what thermal physicists did in the 18th century. Daniel Fahrenheit's scale became popular in the early 1700s because he made and sold especially fine thermoscopes based on measuring the temperature-dependent expansion of mercury in a glass tube. To define his scale, Fahrenheit used a specific mixture of salt and ice to define 0°F (which was about as cold as anyone would experience in daily life in Europe at the time) and defined body temperature to be 96°F (for reasons that need not concern us here). Fahrenheit simply *assumed* that mercury expanded linearly with increasing temperature. The Celsius scale was also based on the mercury thermoscope, except that it defined the freezing and boiling points of water to be 0°C and 100°C, respectively. These reference points proved to be more precise than Fahrenheit's, so this scale ultimately became more popular in scientific circles in the late 1700s.

However, by the early 1800s, the fact that mercury thermoscopes disagreed with those based on the expansion of other liquids or gases called into question whether any could serve as a "true" thermometer. In the 1840s, careful work by Henri Regnault established that thermoscopes based on measuring the pressure (the force magnitude per unit area) exerted by a fixed volume of air (see figure T1.6) yielded more consistently reproducible results than other kinds of thermoscopes, but he worried that even an air-based thermoscope might not be perfect. Indeed, later work showed that thermoscopes containing different densities of air yield (slightly) inconsistent results.

To address this problem, Benjamin Thompson (known after his ennoblement in 1892 as Lord Kelvin) in 1848 proposed an intrinsically *physical* way to define temperature that did not require endorsing *any* particular thermoscope. We will see in chapter T3 how Kelvin's definition emerges naturally from statistical physics and the zeroth law of thermodynamics. We will also see in chapter T5 that the gas pressure in a fixed-volume gas thermoscope is indeed directly proportional to Kelvin's definition of temperature (at least in the limit of low gas density).

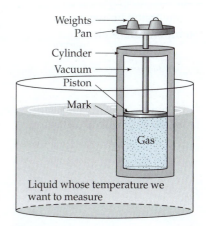

Figure T1.6

An idealized constant-volume gas thermoscope. The gas's pressure (the force per unit area it exerts on the piston) is proportional to the total weight of the piston, pan, and the weights on the pan when the piston is at rest at the mark corresponding to our chosen fixed gas volume.

Kelvin's thermoscope-independent temperature scale

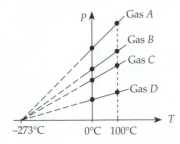

Figure T1.7

The pressures P of confined gases seem to approach zero at a common temperature.

A "thermometer" is a *calibrated* thermoscope

Conversion between temperature scales

Gulliaume Amontons had noted in the early 1700s that the pressure of a fixed volume of air seemed to increase pretty linearly with increasing temperature, and if one projected this linear function back to zero pressure, one seemed to end up at a common temperature, no matter how much gas was involved (see figure T1.7). In 1802, Joseph Gay-Lussac established that this was true for other gases as well, and determined more precisely that the common temperature was −273°C on the Celsius scale. One of the important implications of Kelvin's definition is that this temperature represents the lowest *possible* temperature, as we will see. Kelvin therefore assigned this temperature (which we now call **absolute zero**) the value $T = 0$. With the zero point defined, Kelvin only needed to define the size of an increment on his scale (or, equivalently, the temperature of one other point) to completely define his scale. Because the Celsius scale was widely used in 1848, Kelvin chose his scale's unit to correspond to an increment of 1°C. Kelvin's absolute temperature scale is now the scientific standard, and the SI temperature unit, called the "kelvin" in his honor, is defined in such a way that the freezing point of water has an absolute temperature of $T = 273.15$ K. (Note that the kelvin is never capitalized even though its SI symbol is K. We express a temperature in kelvins, not in "degrees kelvin," and write K, not °K.)

So a **thermometer** is a thermoscope that has been calibrated to be consistent with Kelvin's physical temperature scale. A carefully designed constant-volume gas thermometer can (under the right circumstances) provide a decent empirical way to measure Kelvin's ideal temperature scale.

Because a temperature *difference* of 1 K = 1°C by definition and 0°C corresponds to 273.15 K, we can convert from a temperature T on the kelvin scale to a temperature $T_{[C]}$ on the Celsius scale (or vice versa) by calculating temperature *differences* from a common point (say, water's freezing point):

$$T_{[C]} - 0°C = \frac{1°C}{1\,K}(T - 273.15\,K), \quad T = \frac{1\,K}{1°C}(T_{[C]} - 0°C) + 273.15\,K \quad \text{(T1.2)}$$

The Fahrenheit scale is now *defined* so that a temperature difference of 9°F corresponds to exactly 5 K = 5°C. Moreover, the freezing point of water is 32°F. Therefore, we can convert from one scale to the other as follows:

$$T_{[F]} - 32°F = \frac{9°F}{5°C}(T_{[C]} - 0°C), \quad T_{[C]} - 0°C = T_{[C]} = \frac{5°C}{9°F}(T_{[F]} - 32°F) \quad \text{(T1.3)}$$

Table T1.1 shows selected temperature benchmarks on all three scales.

Table T1.1   Selected temperature benchmarks

Center of the sun	1.5×10^7 K	1.5×10^7°C	2.7×10^7°F
Surface of the sun	5800 K	5500°C	10,000°F
Melting point of tungsten	3683 K	3410°C	6170°F
Melting point of iron	1808 K	1535°C	2795°F
Melting point of lead	601 K	328°C	622°F
Boiling point of water	373 K	100°C	212°F
Normal body temperature	310 K	37°C	98°F
Room temperature	295 K	22°C	72°F
Freezing point of water	273 K	0°C	32°F
Boiling point of nitrogen	77 K	−196°C	−321°F
Boiling point of helium	4.2 K	−269°C	−452°F
Background temperature of universe	2.7 K	−270.5°C	−454.8°F
Lowest laboratory temperatures	<0.1 μK	−273.15°C	−459.7°F

Exercise T1X.3

What is a temperature of −40°F on the Celsius and Kelvin scales?

T1.7   Temperature and Thermal Energy

In unit C, we saw that there was a connection between an object's temperature and the portion of that object's internal energy U that we call *thermal energy* (indeed, from now on we will take U to refer *exclusively* to an object's thermal energy unless otherwise specified). Empirically, we find that under normal circumstances, if we increase an object's thermal energy, its temperature *increases* by an amount given by the equation

Temperature is linked to thermal energy

$$dU = mc\,dT \quad \text{or} \quad c \equiv \frac{1}{m}\frac{dU}{dT} \tag{T1.4}$$

- **Purpose:** This equation describes the empirically observed link between the change dU in an object's thermal energy and its corresponding change in temperature dT, where m is the object's mass, and c is the object's **specific heat** (*not* the speed of light!).
- **Limitations:** Both dU and dT must be small enough that $c \approx$ constant, and the object must maintain the same phase, chemical composition and nuclear composition during the process.
- **Notes:** The right-hand version of this equation basically defines c. Note that *specific heat* is a conventional term, but is a potentially confusing misnomer here, because we can add energy in forms other than heat.

This equation is useful because the specific heat c (so defined) depends almost entirely on the object's *composition*, not (at all) on its mass, and only quite weakly on its temperature.

The fact that c does not depend on mass expresses the idea that the energy dU that we must supply to objects with the same composition to cause a given increase in temperature dT should (logically) be strictly proportional to how much "stuff" the object contains (and thus its mass). Increasing the temperature of a system of two objects with the same mass and composition should, after all, require exactly double the energy as each object separately would require for the same temperature change.

The fact that c only weakly depends on temperature, however, is because we have chosen to use the Kelvin temperature scale. As we will see, the definition of the Kelvin scale means that an object's thermal energy in many circumstances should increase *linearly* with temperature (meaning that $dU/dT =$ constant). Subtleties associated with quantum mechanics and the details of intermolecular interactions do cause deviations from this basic behavior, but at normal temperatures, these deviations are often not large.

Indeed, for certain especially simple kinds of substances (for example, low-density gases and solid objects entirely comprised of a single chemical element, such as a block of iron), we find empirically that the substance's thermal energy per molecule not only increases nearly exactly linearly with temperature but with the constant of proportionality that seems to come very nearly in integer steps:

An intriguing equation for especially simple substances

Table T1.2   dU/dT for various solids and gases

Solid	dU/dT
Lead	$3.18Nk_B$
Gold	$3.06Nk_B$
Silver	$3.00Nk_B$
Copper	$2.95Nk_B$
Iron	$3.01Nk_B$
Aluminum	$2.92Nk_B$

Monatomic gas	dU/dT
Helium	$1.51Nk_B$
Neon	$1.53Nk_B$
Argon	$1.50Nk_B$
Krypton	$1.48Nk_B$
Xenon	$1.51Nk_B$

Diatomic gas	dU/dT
Hydrogen (H_2)	$2.46Nk_B$
Nitrogen (N_2)	$2.50Nk_B$
Oxygen (O_2)	$2.52Nk_B$
Carbon monoxide (CO)	$2.49Nk_B$

Questions to answer

$$dU \approx \left(\frac{n}{2}k_B\right)N\,dT \quad \text{or} \quad \frac{1}{N}\frac{dU}{dT} \approx n\left(\frac{k_B}{2}\right) \tag{T1.5}$$

- **Purpose:** This equation describes the link between the change dU in an object's thermal energy and its change in temperature dT, where N is the number of molecules in the object, n is an integer, and k_B is a universal constant we call **Boltzmann's constant** $= 1.38 \times 10^{-23}$ J/K.
- **Limitations:** This equation applies only to certain simple kinds of substances (monatomic solids and monatomic and diatomic gases), and even then only approximately and over a certain temperature (meaning that dT must be small enough to fit into that range).

For example, at room temperature, $n \approx 6$ (within about 5%) for almost every monatomic solid (a solid consisting of atoms of a single chemical element), $n \approx 3$ (within about 2%) for all low-density monatomic gases (helium, argon, krypton, and so on), and $n \approx 5$ (within about 2%) for low-density diatomic gases (such as oxygen and nitrogen): see table T1.2. This begs for a simple physical explanation, which (as we will see) statistical mechanics provides.

Equation T1.5 suggests a general trend that solids with lighter atoms will have higher specific heats, because such substances will have more atoms per unit mass than those with heavier atoms. For example, iron atoms are about half the mass of cadmium atoms, so we'd expect a kilogram of iron to have roughly double the number of atoms as a kilogram of cadmium, and so require roughly twice the energy to cause a given temperature increase. Indeed, iron's specific heat is 449 J·kg^{-1}K^{-1}, while cadmium's is 230 J·kg^{-1}K^{-1}, which is pretty close to half that of iron. This general principle offers some *qualitative* (though not completely reliable) guidance even for complex substances: for example, granite (which is mostly silicon and oxygen) has a higher specific heat (~800 J·kg^{-1}K^{-1}) than iron, but less than that of paraffin (~2900 J·kg^{-1}K^{-1}), which is mostly carbon and hydrogen.

The idea that objects store thermal energy proved very helpful to us in unit C, allowing us to explain how energy is conserved in situations where it superficially seems to disappear. In our present context, however, this concept raises some perplexing questions:

1. Exactly *how* does an object store energy internally?
2. Why and how is an object's *temperature* linked to its thermal energy? In particular, why does equation T1.5 apply to simple substances?
3. Why does energy flow spontaneously from high to low temperature (as in our paradigmatic thermal process) but never the other way around?

In the next two chapters, we will use Boltzmann's statistical mechanics to explore these questions. By the end of chapter T5, we will see how temperature, energy, and entropy are all linked to the concept of equilibrium, providing full and satisfying answers to these perplexing questions.

TWO-MINUTE PROBLEMS

T1T.1 Characterize each of the following processes as being reversible (A) or irreversible (B). (Some answers may be debatable!)
(a) A living creature grows.
(b) A ball is dropped and falls freely downward.
(c) A ball rebounds elastically from a wall.
(d) A piece of hamburger meat cooks on a grill.
(e) A cube of ice melts in a glass sitting on a table.
(f) A bowling ball elastically scatters some bowling pins.

T1T.2 All irreversible processes involve
(a) macroscopic objects, T or F?
(b) transfers to an object's thermal energy, T or F?
(One or both answers may be debatable!)

T1T.3 In each of the processes described below, energy flows across the boundary of the object that serves as the subject of each sentence. Is this energy flow heat (A), work (B), or some other kind of energy transfer (E)?
(a) The sun emits light into space.
(b) You get cooler when standing in the breeze from a fan.
(c) Your car's brakes get hot when used repeatedly.
(d) Your hands get warm when you rub them together.
(e) Your pizza gets warm in the microwave.

T1T.4 Suppose you stretch a rubber band. The work you do on the rubber band is
A. positive.
B. negative.
C. zero.

T1T.5 A drill bit can become very hot when one is drilling a hole in a metal plate. If we take the system to be the drill bit, does heat flow into or out of the system?
A. Heat flows into the system.
B. Heat flows out of the system.
C. No heat flows in either direction.

T1T.6 Suppose that we place objects A and B into a large bucket of water and allow them to come into equilibrium with the water. If we now extract A and B from the water and immediately place them in contact with each other, they will necessarily be in equilibrium. T or F?

T1T.7 Suppose that we place an aluminum cylinder, a wooden block, and a Styrofoam cup on a table and leave them there for several hours. We then come back into the room and feel each object.
(a) Which (if any) *feels* coolest?
(b) Which (if any) *actually* is coolest?
A. The aluminum cylinder
B. The wooden block
C. The Styrofoam cup
D. All are the same.

T1T.8 Is pressure a (A) *vector* or (B) *scalar* quantity?

T1T.9 Which of the following equations is the correct equation for converting a temperature in kelvins to the equivalent temperature on the Fahrenheit scale?

A. $T_{[F]} = \left(\dfrac{9°F}{1\,K}\right)(T)$

B. $T_{[F]} = \left(\dfrac{9°F}{5\,K}\right)(T - 32°\,F) + 273.15\,K$

C. $T_{[F]} = \left(\dfrac{5\,K}{9°F}\right)(T - 32°\,F) + 273.15\,K$

D. $T_{[F]} = \left(\dfrac{9°F}{5\,K}\right)(T - 273.15\,K) + 32°\,F$

E. $T_{[F]} = \left(\dfrac{9°F}{5\,K}\right)(T - 273.15\,K) - 32°\,F$

F. Other (specify)

T1T.10 Suppose we have a cube containing one mole of gold and another cube containing one mole of aluminum. (The gold cube has more than 7 times the mass of the aluminum cube.) Both are initially at room temperature. Suppose that we increase the temperature of each by 10 K. Which cube's thermal energy likely has increased more?
A. The gold cube
B. The aluminum cube
C. Each cube's internal energy increase is about the same.
D. One cannot say without measuring.

T1T.11 A mole of iron has about half the mass of a mole of silver. Which likely has the greater heat capacity?
A. Iron
B. Silver
C. Both will have about the same heat capacity.
D. One cannot say without measuring.

T1T.12 Suppose that we have two containers, one holding N molecules of helium gas, and one holding $N/2$ molecules of oxygen gas. Both are initially at room temperature. We then add the same amount of energy to each gas. Which is hotter at the end?
A. The helium
B. The oxygen
C. Both have about the same temperature.
D. One cannot say without measuring.

T1T.13 Suppose we place a 100-g aluminum block with an initial temperature of 100°C in a Styrofoam cup containing a 100-g sample of water at 0°C. (The specific heats of aluminum and water are 900 $J \cdot kg^{-1}K^{-1}$ and 4200 $J \cdot kg^{-1}K^{-1}$, respectively.) The system's final temperature is closest to
A. 0°C.
B. 20°C.
C. 50°C.
D. 80°C.
E. 100°C.

HOMEWORK PROBLEMS

Basic Skills

T1B.1   Which of the processes described below are (at least approximately) reversible? Which are irreversible? Explain your reasoning briefly in each case.
(a) A ball rolls down an incline.
(b) You catch a fly ball.
(c) A meteor vaporizes as it streaks across the sky.
(d) A trampoline launches you upward.
(e) You pour a glass of water into a pitcher.

T1B.2   Which of the processes described below are (at least approximately) reversible? Which are irreversible? Explain your reasoning briefly in each case.
(a) A brick lands with a thud on the ground.
(b) A pendulum swings from the left to the right.
(c) You stir a spoonful of cocoa into a glass of milk.
(d) A baseball and basketball collide in midair.
(e) You fold a sheet of paper.

T1B.3   Answer the questions posed in problem T1T.3, explaining your reasoning in each case.

T1B.4   In each of the processes described below, energy flows across the boundary of the object that serves as the subject of each sentence. Is this energy flow heat, work, or some other kind of energy transfer? Briefly explain your reasoning in each case.
(a) A car skids to a stop, causing its tires to smoke.
(b) A chunk of metal is vaporized by a laser.
(c) A ball deforms as you squeeze it.
(d) Your coffee cools as it sits in a cup on your table.
(e) Air in a cylinder gets colder as you allow the cylinder's volume to increase.

T1B.5   In figure T1.6, how much weight will we have to put on the pan if the pressure of the gas is $10,000 \text{ N/m}^2$ and the area of the piston is 3.0 cm^2? Assume that the piston and pan assembly itself has a mass of 20 g.

T1B.6   Following the method that we used to arrive at equations T1.2 and T1.3, find an equation for converting a temperature on the Fahrenheit scale to the equivalent temperature in kelvins.

T1B.7   The Rankine temperature scale [where temperatures are expressed in degrees Rankine (°R)] is defined so that temperature differences in °R are the same as those in °F, but the zero of the Rankine scale is at absolute zero.
(a) Find a conversion formula between temperatures on the Rankine scale and those on the kelvin scale.
(b) What is room temperature on the Rankine scale?

T1B.8   The hottest recorded daytime temperature on earth is about 130°F. What is this temperature on the Celsius scale? In kelvins?

T1B.9   The lowest officially recorded temperature within the continental United States is about −70°F. What is this temperature on the Celsius scale? In kelvins?

T1B.10   Body temperature is often quoted as being 98.6°F. However, a given person's body temperatures vary over a range of as much as ±0.9°F during a day, and can vary by about as much from person to person. So it seems strange to quote this temperature to a tenth of a degree: it would make more sense to round it to the nearest integer and say that it is 99°F. However, convert 98.6°F to the equivalent temperature on the Celsius scale. Given that *most* European scientists in the 19th century were using the Celsius scale, speculate on why body temperature is stated as 98.6°F instead of 99°F in countries using the Fahrenheit scale.

T1B.11   About how much energy do we need to increase a mole of helium's temperature by 10 K?

T1B.12   About how much energy do we need to increase a half-mole of nitrogen's temperature by 40 K?

Modeling

T1M.1   Your lab partner claims that physicists have it all backward: cold things actually have more thermal energy than hot things, and energy actually flows from cold to hot. What evidence could you point out that would contradict this assertion?

T1M.2   *Must* we represent hotter temperatures by higher numbers, or is this just a convention? If it is not a convention, explain why hotter temperatures *necessarily* must be represented by higher numbers. If this is a convention, suggest why people might have chosen this convention.

T1M.3   Suppose we measure the pressure of the gas in a constant-volume gas thermoscope to be $32,000 \text{ N/m}^2$ at the freezing point of water and $42,300 \text{ N/m}^2$ when immersed in a given liquid. What is the liquid's temperature?

T1M.4   Suppose we measure the pressure of the gas in a constant-volume gas thermoscope to be $55,000 \text{ N/m}^2$ at the freezing point of water and $42,300 \text{ N/m}^2$ when immersed in a given liquid. What is the liquid's temperature?

T1M.5   Suppose you use a microwave oven to increase the temperature of a cup (about 250 g) of water to boiling. Water has a specific heat of about $4200 \text{ J·kg}^{-1}\text{K}^{-1}$.
(a) While the microwave is running, is heat flowing into or out of the water? Explain.
(b) When the water in the cup is boiling, you remove it from the microwave and place it on the counter. When you come back to it two minutes later, you find that the water's temperature has cooled to 85°C. At about what rate (in watts) has heat flowed out of the water?

T1M.6 Suppose we put a 100-g block of aluminum with an initial temperature of 100°C into a cup containing 250 g of water at 25°C. What is the final equilibrium temperature of this system? (*Hint:* You may use the result of problem T1D.2. The specific heats of aluminum and water are about 900 J·kg^{-1}K^{-1}, and 4200 J·kg^{-1}K^{-1}, respectively.)

T1M.7 Suppose we put a 150-g steel ball with an initial temperature of 0°C into a cup containing 150 g of water with an initial temperature of 100°C. What is the final equilibrium temperature of this system? (*Hint:* You may use the result of problem T1D.2. The specific heats of iron and water are 450 J·kg^{-1}K^{-1} and 4200 J·kg^{-1}K^{-1}, respectively.)

T1M.8 The atomic weight (mass per mole) of silver is about 2.25 times that of titanium, which has a specific heat of about 523 J·kg^{-1}K^{-1}. Predict the specific heat of silver, and explain your reasoning. (You might check your reasoning by looking up silver's specific heat.)

T1M.9 The atomic weight (mass per mole) of tin is about 2.02 times that of nickel, which has a specific heat of about 445 J·kg^{-1}K^{-1}. Predict the specific heat of tin and explain your reasoning. (You might check your reasoning by looking up tin's specific heat.)

Derivation

T1D.1 If equation T1.5 were to hold true all the way to absolute zero (it doesn't), how would U depend on T?

T1D.2 Suppose we place object A with high initial temperature T_A in contact with object B with low initial temperature T_B. We know from experience that thermal energy will flow from A to B under these circumstances. If these objects are isolated from everything else during this process, then any energy lost by A will be gained by B:

$$dU_A = -dU_B \qquad (T1.6)$$

(a) If this change of thermal energy is "sufficiently small" that the objects' specific heats are approximately constant over the temperature ranges involved, then show, using this equation and equation T1.4, that

$$-\frac{dT_B}{dT_A} = \frac{m_A c_A}{m_B c_B} \qquad (T1.7)$$

(b) Now, as both objects eventually come to the same final temperature T_f, the total change in the temperature of A is $dT_A = T_f - T_A$ and that for B is $dT_B = T_f - T_B$. To save writing, let $u \equiv m_A c_A/m_B c_B$. Plug these relations into equation T1.7, and show (after some algebra) that

$$T_f = T_B + \frac{u}{1+u}(T_A - T_B) \qquad (T1.8)$$

(c) Note that if object A is much more massive than object B, then u will be very large and $u/(1+u) \approx 1$, meaning that $T_f \approx T_B + T_A - T_B$, as one might expect. Argue that equation T1.8 also gives a plausible result if object B is much more massive than object A.

T1D.3 Who cares how we define temperature? Suppose that scientists on the planet Serendipa happened to choose as the temperature standard a thermoscope whose temperature scale is related to the Kelvin scale by the formula $T_S = \alpha T^2$, where T_S is the temperature in Serendipa units, T is the temperature in kelvins, and α is some constant.
(a) Does this scale satisfy the zeroth law?
(b) Show that the quantity $(1/N)(dU/dT_S)$ is *not* constant for simple gases and/or solids, as it is when we use the Kelvin scale.
(c) Might the scientists on Serendipa miss seeing something valuable if they used this scale? Discuss.

T1D.4 **(a)** Derive an approximate expression for the specific heat c of an elemental solid (that is, a solid consisting of a pure chemical element) in terms of k_B and the atomic weight (mass per mole) M_A of the element.
(b) Check your formula by predicting the specific heats of iron, copper, and gold, and compare your predictions with their measured specific heats (450 J·kg^{-1}K^{-1}, 385 J·kg^{-1}K^{-1}, and 129 J·kg^{-1}K^{-1}, respectively).

Rich-Context

T1R.1 Suppose that you have three constant-volume gas thermoscopes of the type shown in figure T1.6. When you put these thermoscopes into ice water, their pressures are 11.3 kN/m^2, 18.3 kN/m^2, and 26.5 kN/m^2, respectively. When you put them into boiling water, their pressures are 14.5 kN/m^2, 25 kN/m^2, and 36.2 kN/m^2, respectively.
(a) One of these thermoscopes is bad (perhaps because it is leaking gas). Which one and how do you know?
(b) Calibrate the other two thermoscopes by finding the conversion factor between the pressure of the gas in that thermoscope and the temperature in kelvins.

T1R.2 Suppose you are given a 4.8-g hunk of an unidentified shiny, silvery metal that you know is from a collection of samples of pure chemical elements, but you do not know the element. You find that it takes almost exactly 2.49 J of energy to increase its temperature by 1.00 K. Using only information in this text and a periodic table, identify the most likely metal and explain your method.

ANSWERS TO EXERCISES

T1X.1 Reversible: (b) and (e); irreversible: (a), (c), and (d)

T1X.2 **(a)** Q, **(b)** Q, **(c)** W, **(d)** Q, **(e)** W, **(f)** Q, **(g)** Q, **(h)** $[E]$, **(i)** W.

T1X.3 $-40°F = -40°C = 233$ K

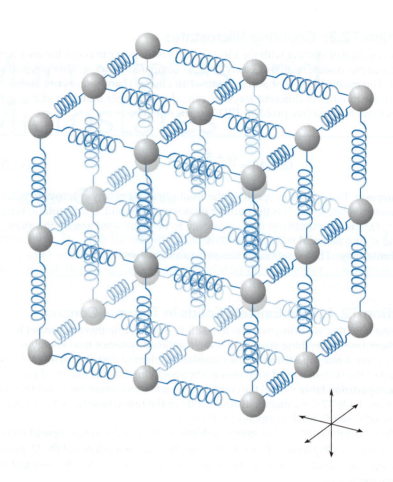

Figure T2.1

We can model the interactions between atoms in a crystalline monatomic substance as springs.

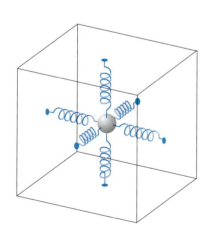

Figure T2.2

In Einstein's model, we treat each atom as independent, as if its springs were connected to the rigid walls of its "cell."

T2.1   The Einstein Model of a Solid

In this chapter, we begin exploring Boltzmann's solution to the puzzle of irreversibility. We will be able to understand this solution most easily with the help of a very simple model of a monatomic solid.

In 1907, Albert Einstein published a paper that proposed a simple but reasonably accurate model for predicting the thermal behavior of monatomic crystalline solids (such as crystals of pure carbon, iron, or gold). Atoms in such a solid are held in nearly fixed positions in the crystal lattice by complicated interatomic electromagnetic interactions. However, as long as the atoms remain close to their equilibrium positions (where their interaction potential energy functions are minimum), we saw in chapter C9 that we can approximate the potential energy function for almost any interaction by that for a spring, so we can (as a first approximation) model all the interatomic interactions as if they were springs (see figure T2.1).

Even in such a model, the atoms can affect each other in complicated ways, so Einstein proposed a further simplification: assume that each atom oscillates *independently* about its equilibrium position, behaving as if it were connected by springs to the rigid walls of its individual "cell" in the crystal lattice instead of being interconnected with the surrounding atoms (see figure T2.2). So in this model, we assume that each atom's potential energy when it is a distance r from its equilibrium position is $\frac{1}{2}k_s r^2$ (where k_s is the effective spring constant of the interactions that hold the atom in place) independent of both the direction of r and the positions of neighboring atoms. The model also assumes that each atom is identical, meaning that the spring constant k_s is the same for all atoms. Einstein was able to show (as we will see) that this simplistic model nonetheless accurately describes the thermal behavior of monatomic solids over a wide temperature range.

Now, in both Newtonian and quantum mechanics, we can treat a particle oscillating in three dimensions as if it were three independent *one*-dimensional oscillators. For example, the total Newtonian energy of a three-dimensional oscillator is

We can treat a three-dimensional oscillator as three one-dimensional oscillators

$$E = \tfrac{1}{2}m|\vec{v}|^2 + \tfrac{1}{2}k_s r^2 = \tfrac{1}{2}m(v_x^2 + v_y^2 + v_z^2) + \tfrac{1}{2}k_s(x^2 + y^2 + z^2)$$
$$= (\tfrac{1}{2}mv_x^2 + \tfrac{1}{2}k_s x^2) + (\tfrac{1}{2}mv_y^2 + \tfrac{1}{2}k_s y^2) + (\tfrac{1}{2}mv_z^2 + \tfrac{1}{2}k_s z^2) \qquad \text{(T2.1)}$$

where m is the atom's mass. Note how we can group the terms in the energy equation into three pairs, each pair of which would be the energy associated with a *one*-dimensional oscillation along one of the coordinate axes, without any reference to what is happening in the other coordinate directions. One can also show from this equation (see problem T2D.5) that the atom's motion along each coordinate axis is exactly as if the atom were oscillating in one dimension along that axis *alone*. (This is a special property of the spring potential energy function: most other potential energy functions cannot be pulled apart in this way.)

According to quantum mechanics (see chapter Q10), the energy associated with each of these separate one-dimensional oscillations is *quantized*, so that the atom's total vibrational energy is given by

$$E = \hbar\omega(n_x + \tfrac{1}{2}) + \hbar\omega(n_y + \tfrac{1}{2}) + \hbar\omega(n_z + \tfrac{1}{2}) \qquad \text{(T2.2)}$$

where $\omega = (k_s/m)^{1/2}$ is the angular frequency of the equivalent Newtonian oscillator and $\hbar \equiv h/2\pi$ where h is Planck's constant. The quantities n_x, n_y, and n_z here are independent, nonnegative integers (0, 1, 2, 3, and so on) that specify each oscillator's energy level. Each of the three terms here is the same as the expression for the energy of a *one*-dimensional quantum harmonic oscillator. See unit Q for a justification of this: for our purposes at the present it is enough simply to know that this is true.

We can rewrite this equation as

$$E = \sum_{i=1}^{3} \hbar\omega(n_i + \tfrac{1}{2}) = \sum_{i=1}^{3} \varepsilon(n_i + \tfrac{1}{2}) \qquad \text{(T2.3)}$$

where $\varepsilon \equiv \hbar\omega = \hbar(k_s/m)^{1/2}$ is the energy difference between adjacent levels of each one-dimensional oscillator and n_1, n_2, and n_3 are just a different way of labeling the integers n_x, n_y, and n_z. We can then find the solid's total energy by summing this over all atoms (three terms per atom):

$$E_{\text{tot}} = \sum_{i=1}^{3N} \varepsilon\left(n_i + \frac{1}{2}\right) = \sum_{i=1}^{3N} \varepsilon n_i + \sum_{i=1}^{3N} \frac{1}{2}\varepsilon = \sum_{i=1}^{3N} \varepsilon n_i + \frac{3}{2}N\varepsilon \qquad \text{(T2.4)}$$

The constant $\frac{3}{2}N\varepsilon$ term in this equation is called the solid's **zero-point energy**. The solid will have this energy even at absolute zero (which is the temperature where all atoms are in their *lowest possible* energy state, by definition) and cannot be affected by the solid's interaction with its surroundings: it is a built-in aspect of the solid's internal energy. We define a system's *thermal energy* U to be that part of the system's internal energy that *changes* when the system's temperature changes due to its interactions with other objects (assuming that the system's phase, chemical composition, and nuclear composition remain fixed). So, this zero-point energy (along with the internal energies of all the atoms) is not part of the solid's *thermal* energy U. We can therefore write the solid's internal energy as

The quantum zero-point energy of the solid is irrelevant

The thermal energy of an Einstein solid

$$U = \sum_{i=1}^{3N} \varepsilon n_i \quad \text{where} \quad \varepsilon = \hbar \sqrt{\frac{k_s}{m}} \qquad \text{(T2.5)}$$

- **Purpose:** This equation specifies the total thermal energy U of an Einstein solid whose N atoms we model as $3N$ identical but independent quantum oscillators, where $n_1, n_2, n_3, \ldots$ are a set of nonnegative integers (three per atom), ε is the fixed difference between each oscillator's energy levels, $\hbar \equiv h/2\pi$, h is Planck's constant, k_s is the effective spring constant of the interactions holding an atom in place, and m is the mass of one of the solid's atoms.
- **Limitations:** The Einstein model works well for monatomic solids at temperatures above 100 K or so (the exact limit depends on the solid).

Note that because $\varepsilon \propto (k_s/m)^{1/2}$, it increases as the strength of the interatomic forces increases, and decreases as the mass of each of the atoms increases.

The basic point of equation T2.5 is that *we will model a crystalline solid containing N identical atoms as if it contained 3N identical independent quantum harmonic oscillators, each of which can store an integer number n_i of energy units ε.* We will call any solid accurately described by this model an **Einstein solid.**

T2.2   Distinguishing Macrostates and Microstates

The core step in understanding Boltzmann's solution to the problem of irreversibility is to understand the crucial distinction between the *macrostate* and the *microstate* of a thermodynamic system.

Definition of *macrostate*

A system's **macrostate** is that system's thermodynamic state as characterized by its *macroscopically measurable and potentially variable properties.* These are properties of the system *as a whole* that might in principle change as the system interacts thermally with its surroundings. These properties include the system's total thermal energy U, its temperature T, its volume V, the number of particles N that it contains, its mass M, and so on.

We can completely describe a system's macrostate by specifying some *minimal* set of macroscopic properties that suffice to calculate all its *other* macroscopic properties. For example, a monatomic solid's thermal energy U is connected to its temperature T and the number of particles N (see equation T1.5), so if we specify U and N, we can in principle calculate T. The solid's mass $M = mN$, where m is the mass of an atom. As the atom's mass does not depend on the system's external circumstances, it does not count as a *variable* property, so knowing N determines M. Under normal circumstances, a solid's density is also pretty much fixed, so knowing N also determines the solid's volume V. We see, therefore, that knowing U and N pretty much fixes a solid's macrostate under normal circumstances. On the other hand, the density of a gas (or a solid under extreme pressure) might vary dramatically, so we might have to add another macroscopic variable (for example, the system's volume V) to completely describe such a system's macrostate.

Definition of *microstate*

A system's **microstate,** on the other hand, is characterized by describing the quantum state of *each individual molecule* in the system at a given time. In the case of an Einstein solid, this means specifying the quantum state of (that is, the value of n for) each of the system's $3N$ independent oscillators. Note that while this idea is conceptually straightforward, even the tiniest speck of solid contains so many atoms that describing the speck's microstate would be impossible in practice, but we can at least *imagine* doing it.

Note that if we know a system's microstate, we know its macroscopic properties as well. If we know the value of n for every oscillator in an Einstein solid, then we implicitly know N and can calculate U using equation T2.5, so we know what we need to specify the solid's macrostate.

Now, the most important thing to understand about microstates and macrostates is that a system in a given, well-specified macrostate could be in any one of a *huge* number of different microstates that we are unable to distinguish by macroscopic measurements. For example, suppose that we describe an Einstein solid's macrostate by stating values for U and N. There are many possible microstates that nonetheless add up to the same total U (each simply corresponds to a different way of distributing that total energy among $3N$ oscillators). For any complex system, there are an *immense* number of possible microstates consistent with any given macrostate.

Perhaps the following analogy will make these ideas more vivid. Consider your bedroom. It has two fundamental "macrostates" that a person (say, your parent) can rapidly discern without much detailed examination: "clean" or "messy." Describing your room's microstate, on the other hand, would involve meticulous documentation of the exact position and orientation of every object in the room. Now, a fairly large number of arrangements of objects in your room might qualify the room as being "clean" (for example, there are a number of possible ways to neatly arrange your socks in the dresser drawer). There are vastly *more* possibilities for object arrangements that your parent would consider messy (just imagine the number of ways you could distribute your socks on the floor!). Either way, though, there are many microstates in a macrostate.

There are *many* microstates in a macrostate

T2.3 Counting Microstates

Why begin our journey with the Einstein solid? The answer is that we can *much* more easily calculate the number of microstates corresponding to each macrostate of an Einstein solid than for any other reasonably realistic thermodynamic model. This in turn makes it comparatively easier to determine what statistical physics predicts about this model. Models of other complex systems behave in *qualitatively* similar ways, so what we learn from this model applies at least qualitatively to other systems as well. We will also use this model to develop tools that help us handle more complex models.

As we saw in the last section, we can describe an Einstein solid's macrostate by specifying its thermal energy U and its number of atoms N. To describe the solid's microstate, we must specify an integer value n_i for each of the solid's $3N$ independent oscillators. In general, there will be *many* microstates (that is, many distinct sets of values for all $3N$ integers n_i) that have the same total U. We call the number of possible microstates that correspond to the same given macrostate that macrostate's **multiplicity** Ω. In the case of an Einstein solid, where the macrostate is specified by U and N,

Describing the macrostate of an Einstein solid

$\Omega(U, N) \equiv$ the multiplicity of the macrostate specified by U and N
$\qquad =$ the number of N-atom microstates having total energy U (T2.6)

How can we determine $\Omega(U, N)$ for given values of U and N? The beauty of the Einstein solid model is that this is not a *conceptually* difficult problem. According to equation T2.5, the total energy in an Einstein solid is an integer multiple of the basic energy unit ε. Think of each energy unit as a marble, and each of the solid's $3N$ oscillators as a bin into which we can put marbles. When we specify the solid's total energy U, we are essentially specifying the

Counting microstates

total number of "marbles" $q \equiv U/\varepsilon$ that we must distribute. Counting the microstates for this U, then, is the same as counting how many different ways we can sort q marbles into $3N$ bins.

In a solid of macroscopic size, $3N$ will be on the order of magnitude of 10^{23}, but let's start small. Suppose our Einstein solid consists of a *single* atom, that is, *three* independent oscillators. (Of course, a single atom will have no lattice in which it can oscillate, but let us just pretend that this makes sense. We'll work up to larger numbers of atoms shortly.) We can describe the microstate of this system by specifying energy-level integers for each of the three oscillators (that is, how many marbles each of these three bins contains). Let us write these numbers as a triplet of digits; for example, the triplet 032 specifies the microstate in which the first oscillator has 0 units of energy, the second has 3 units, and the third has 2 units. The total energy contained in the system in this case is $U = 5\varepsilon$. Other possible microstates corresponding to this macrostate are 320, 230, 302, 203, 023, 113, 311, 131, 041, 014, and so on.

So let us start counting microstates for various different macrostates of this one-atom Einstein solid. First, suppose the solid's total energy has its lowest possible value $U = 0$. Only one microstate (000) is compatible with this total energy, so this macrostate's multiplicity is $\Omega(U, N) = \Omega(0, 1) = 1$.

Now suppose the solid's total energy is $U = \varepsilon$ (that is, the solid contains exactly 1 unit of energy). The microstates compatible with this total energy are 100, 010, and 001, for a total of three. This macrostate's multiplicity is thus $\Omega(1\varepsilon, 1) = 3$. If the solid's total energy is $U = 2\varepsilon$, then the possible microstates are 002, 020, 200, 110, 101, 011, so this macrostate's multiplicity is $\Omega(2\varepsilon, 1) = 6$. In a similar fashion, one can show that $\Omega(3\varepsilon, 1) = 10$, $\Omega(4\varepsilon, 1) = 15$, $\Omega(5\varepsilon, 1) = 21$, and $\Omega(6\varepsilon, 1) = 28$.

Exercise T2X.1

Verify that $\Omega(3\varepsilon, 1) = 10$ by writing down all possible microstate triplets consistent with this macrostate and counting them.

An Einstein solid with two atoms (six independent oscillators) and zero total energy has one microstate 000000, so $\Omega(0, 2) = 1$. If the solid's total energy is $U = \varepsilon$, then the possible microstates are 000001, 000010, 000100, 001000, 010000, and 100000, so $\Omega(1\varepsilon, 2) = 6$. When $U = 2\varepsilon$, the possible microstates are 000002, 000020, 000200, 002000, 020000, and 200000, 000011, 000101, 001001, 010001, 100001, 000110, 001010, 001100, 010010, 010100, 011000, 100010, 100100, 101000, and 110000, so $\Omega(2\varepsilon, 2) = 21$. In a similar fashion, one can show that $\Omega(3\varepsilon, 2) = 56$, $\Omega(4\varepsilon, 2) = 126$, and so on.

Counting microstates this way gets pretty tedious after a while. However, one can prove generally that if $q = U/\varepsilon$ is the total number of energy units to be distributed among $3N$ oscillators, then

$$\Omega(U, N) = \frac{(q + 3N - 1)!}{q!(3N - 1)!} \tag{T2.7}$$

- **Purpose:** This equation specifies the multiplicity Ω of any macrostate of an Einstein solid, where N is the number of atoms in the solid, U is its total energy, $q \equiv U/\varepsilon$ is the number of units of energy to be distributed among the atoms, and $n!$ or **n factorial** $\equiv 1 \cdot 2 \cdot 3 \cdots (n - 1) \cdot n$.
- **Limitations:** This equation applies *only* to an Einstein solid.

(Problem T2D.3 shows how to derive this formula.) So, for example,

$$\Omega(4\varepsilon, 2) = \frac{(4 + 6 - 1)!}{4!(6 - 1)!} = \frac{9!}{4!5!} = \frac{9 \cdot 8 \cdot 7 \cdot 6 \cdot 5 \cdot 4 \cdot 3 \cdot 2 \cdot 1}{(4 \cdot 3 \cdot 2 \cdot 1)(5 \cdot 4 \cdot 3 \cdot 2 \cdot 1)}$$

$$= \frac{9 \cdot 8 \cdot 7 \cdot 6}{4 \cdot 3 \cdot 2 \cdot 1} = 9 \cdot 7 \cdot 2 = 126 \qquad \text{(T2.8)}$$

Exercise T2X.2

(a) Check that equation T2.7 yields the same results for $\Omega(0, 1)$, $\Omega(\varepsilon, 1)$, $\Omega(2\varepsilon, 1)$, and $\Omega(2\varepsilon, 2)$ that we found earlier by direct counting. **(b)** Use equation T2.7 to verify that $\Omega(6\varepsilon, 1) = 28$. **(c)** If an Einstein solid has *three* atoms, what is the multiplicity of the macrostate where it has 8 units of energy?

T2.4 Two Einstein Solids in Thermal Contact

Suppose now we bring two Einstein solids A and B, one with N_A atoms ($3N_A$ oscillators) and one with N_B atoms ($3N_B$ oscillators), into thermal contact, so that microscopic interactions between atoms on the surfaces in contact can allow energy to flow between the solids. How will these solids behave *macroscopically* after being brought into contact?

The values of N_A and U_A specify solid A's macrostate and N_B and U_B. specify solid B's macrostate. If we suppose that N_A and N_B are fixed, then just U_A and U_B sufficiently specify the solids' macrostates. If the combined system of the two solids is thermally isolated, its total energy $U = U_A + U_B$ is fixed (by conservation of energy); but at least in principle, the energies U_A and U_B of the two solids could have any values consistent with that total. For example, if the combined system's total energy is $U = 6\varepsilon$, then possible pairs of values for U_A and U_B include $U_A = 0$ and $U_B = 6\varepsilon$, or $U_A = 2\varepsilon$ and $U_B = 4\varepsilon$, or $U_A = 5\varepsilon$ and $U_B = \varepsilon$, and so on.

Let' call a given pair of macrostates for solids A and B that are consistent with a fixed value of $U = U_A + U_B$ a **macropartition** of the combined system for that U. For example, the pair of macrostates where $U_A = 2\varepsilon$ and $U_B = 4\varepsilon$ is one possible *macropartition* of the combined system for $U = 6\varepsilon$.

The *macropartition* of a pair of objects in thermal contact

Different macropartitions of the combined system of two solids therefore amount to different ways that the energy can be *macroscopically* divided (or "partitioned") between the solids. There is a real distinction to be made here between a *macropartition* and a *microstate* of the combined system. A microstate of the combined system specifies exactly how much energy *each individual oscillator* in both solids has. A macropartition, on the other hand, only specifies the *macroscopic* total energies U_A and U_B that the two macroscopic solids have, something we can measure macroscopically. In other words, we describe a macropartition of a combined system of two subsystems by describing *the macrostate of each subsystem*.

Now, suppose that in a certain macropartition, solid A has energy U_A and multiplicity Ω_A, and solid B has energy U_B and multiplicity Ω_B. What is the macropartitition's multiplicity? Well, for *each* of the Ω_A microstates that solid A might be in, solid B could be in *any* of its Ω_B microstates, so the total number of microstates consistent with this particular partitioning of the energy between solids A and B must be the *product* of these multiplicities:

A macropartition's multiplicity

$$\Omega_{AB} = \Omega_A \Omega_B \qquad \text{(T2.9)}$$

A macropartition table for Einstein solids in contact

Table T2.1   Table of possible macropartitions for $N_A = N_B = 1$, $U = 6\varepsilon$

U_A	U_B	Ω_A	Ω_B	Ω_{AB}
0	6	1	28	28
1	5	3	21	63
2	4	6	15	90
3	3	10	10	100
4	2	15	6	90
5	1	21	3	63
6	0	28	1	28

Total microstates = 462

A specific example may make this clearer. Suppose that we bring two one-atom Einstein solids into thermal contact, and suppose that their total combined energy is $U = 6\varepsilon$. Let's construct a table that lists in successive lines all the possible macropartitions for the combined system. On each line, we specify the macropartition by stating the two solids' energies U_A and U_B in units of ε, the multiplicities Ω_A and Ω_B of their respective macrostates, and the macropartition's multiplicity $\Omega_{AB} = \Omega_A\Omega_B$. Table T2.1 shows such a **macropartition table.**

To see why $\Omega_{AB} = \Omega_A\Omega_B$, consider, for example, the macropartition where $U_A = U_B = 3$. The possible microstates of solid A are (in our previous notation) 300, 030, 003, 210, 201, 021, 120, 102, 012, and 111, and the possible microstates of system B are the same. The possible microstates of the combined system are as follows (the triplets on the left and right specify the microstates of solids A and B, respectively): 300-300, 300-030, 300-003, 300-210, 300-201, 300-021, 300-120, 300-102, 300-012, 300-111, 030-300, 030-030, 030-003, 030-210, and so on, for a total of $10 \times 10 = 100$ distinct microstates.

Exercise T2X.3

Prepare an analogous table for the case where $N_A = N_B = 1$ and $U = 8\varepsilon$. (You can copy most of the multiplicities Ω_A and Ω_B from table T2.1; use equation T2.7 to calculate the rest.)

T2.5   The Fundamental Assumption

In a real solid, the atoms are not quite *completely* independent, they can (and do) exchange energy with each other through random microscopic processes. So to get the Einstein model to fit reality, we must assume that adjacent atoms *do* interact enough to exchange energy, but not *so* strongly that the energy-level structure of each quantum oscillator is significantly affected. Energy will also shift randomly back and forth across the boundary between the solids by the interactions between atoms on the surfaces in contact.

Therefore, as time passes, a combined system of two Einstein solids will randomly shift between different microstates consistent with the constraint that the total energy has some fixed value U. This means that under some circumstances, the macropartition of the combined system might fluctuate as the system randomly samples microstates in different macropartitions. For example, in the situation considered in table T2.1, the combined system in microstate 012-300 (one of the microstates corresponding to the macropartition where $U_A = U_B = 3$) might evolve to 013-200 (one of the microstates corresponding to macropartition $U_A = 4$, $U_B = 2$) by transferring 1 unit of energy across the boundary. In time, this system will sample each of the 462 possible microstates, and thus each of the possible macropartitions.

Now comes the big question: Can we say something about which macropartitions we are most *likely* to see if we peek at the system at various times? We can indeed, if we are willing to accept a simple and plausible assumption:

The fundamental assumption of statistical mechanics

> An isolated system's accessible microstates are all *equally likely* in the long run.

(*Accessible* in this context means "consistent with the value of the total internal energy of the system in question.") We call this statement the **fundamental assumption** of statistical mechanics.

This disarmingly simple postulate provides the foundation for under-standing irreversible processes, as we will shortly see. Note that even though this assumption is simple and plausible, its ultimate justification is that it does correctly predict the behavior of macroscopic systems.

The most important consequence of this principle is that the probability that a given energy macropartition (consistent with the given total internal energy) will occur is directly proportional to the number of microstates that indistinguishably generate that macropartition; that is, to that macroparti-tion's total multiplicity Ω_{AB}.

For example, suppose we were to take a large number of "snapshots" of the system of two Einstein solids described in table T2.1. The fundamental assumption means, over the long haul, we should find the system to be in macropartition 3:3 in about $100/462 = 0.216 = 21.6\%$ of the pictures and in macropartition 0:6 (or macropartition 6:0) in about $28/462 = 0.061 = 6.1\%$ of the pictures, and so on. Note that the macropartition $U_A = U_B = 3$, the mac-ropartition for which the energy is shared equally between the two identical solids, is the system's single most probable macropartition.

T2.6   Using StatMech

Doing the calculations required to set up a table such as table T2.1 can be quite tedious, particularly as the number of atoms in each solid becomes large. Fortunately, you can run a web application (called StatMech) that does these calculations for you. You can access StatMech by going to the web page

        http://physapps.pomona.edu/

When you run the application, it initially displays a modified version of table T2.1 (see figure T2.3): the major difference is that the table entries also display U_A/U and U_B/U in the third and fourth columns and the macro-state's probability in the final column. StatMech also displays a graph of the probability as a function of U_A/U (not shown in figure T2.3).

You can generate a new table (and corresponding graph) by entering new values for N_A, N_B, and U into the fields on the left. After you enter the values, press the "Update" button to display the new table and graph. (Note that you need to press "Update" after *any* change in the values of text boxes or controls on the window to see the results.)

In principle, you can do *by hand* what StatMech does, at least when N_A, N_B, and U are reasonably small (indeed, you did this in exercise T2X.3). Once you have constructed a few tables on your own, however, you will appreciate how rapidly and easily the application generates tables.

You can enter fairly large numbers for N_A and N_B without any problem. If you enter numbers for U that are larger than 100, each row of the table will display the summed results for a "bin" of macropartitions instead of the results for an individual macropartition. The application does this so that the

Basic instructions for using StatMech

Figure T2.3

StatMech's initial table.

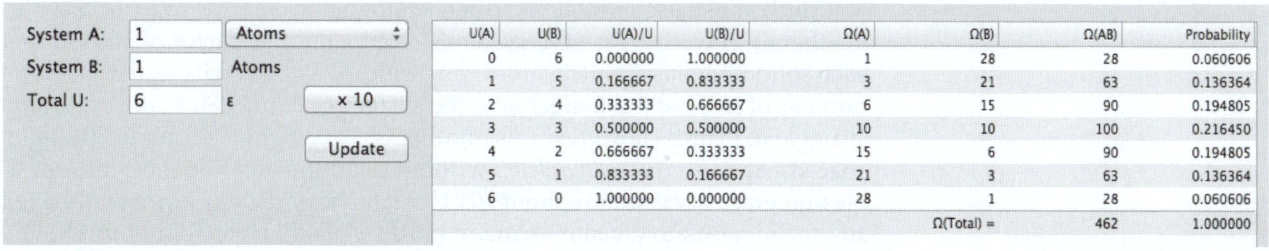

System A:	1	Atoms	⬍
System B:	1	Atoms	
Total U:	6	ε	× 10
			Update

U(A)	U(B)	U(A)/U	U(B)/U	Ω(A)	Ω(B)	Ω(AB)	Probability
0	6	0.000000	1.000000	1	28	28	0.060606
1	5	0.166667	0.833333	3	21	63	0.136364
2	4	0.333333	0.666667	6	15	90	0.194805
3	3	0.500000	0.500000	10	10	100	0.216450
4	2	0.666667	0.333333	15	6	90	0.194805
5	1	0.833333	0.166667	21	3	63	0.136364
6	0	1.000000	0.000000	28	1	28	0.060606
					Ω(Total) =	462	1.000000

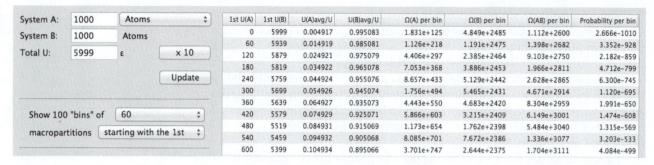

1st U(A)	1st U(B)	U(A)avg/U	U(B)avg/U	Ω(A) per bin	Ω(B) per bin	Ω(AB) per bin	Probability per bin
0	5999	0.004917	0.995083	1.831e+125	4.849e+2485	1.112e+2600	2.666e-1010
60	5939	0.014919	0.985081	1.126e+218	1.191e+2475	1.398e+2682	3.352e-928
120	5879	0.024921	0.975079	4.406e+297	2.385e+2464	9.103e+2750	2.182e-859
180	5819	0.034922	0.965078	7.053e+368	3.886e+2453	1.966e+2811	4.712e-799
240	5759	0.044924	0.955076	8.657e+433	5.129e+2442	2.628e+2865	6.300e-745
300	5699	0.054926	0.945074	1.756e+494	5.465e+2431	4.671e+2914	1.120e-695
360	5639	0.064927	0.935073	4.443e+550	4.683e+2420	8.304e+2959	1.991e-650
420	5579	0.074929	0.925071	5.866e+603	3.215e+2409	6.149e+3001	1.474e-608
480	5519	0.084931	0.915069	1.173e+654	1.762e+2398	5.484e+3040	1.315e-569
540	5459	0.094932	0.905068	8.085e+701	7.672e+2386	1.336e+3077	3.203e-533
600	5399	0.104934	0.895066	3.701e+747	2.644e+2375	1.704e+3111	4.084e-499

Figure T2.4

What StatMech's macropartition table looks like when $U > 100$.

How StatMech describes "bins" of macropartitions

number of table rows does not become too unwieldy. (The current maximum limit on U is 999999ε to keep computation time reasonable, though this may increase as computers become faster.)

As figure T2.4 shows, the table layout becomes somewhat different when the macropartitions are grouped into "bins": the first and second columns specify U_A and U_B for the *first* macropartition in each bin, the third and fourth columns specify the *average* values of the fractional energies U_A/U and U_B/U for the macropartitions in the bin, and the remaining columns specify the *total* multiplicities and the probability of each bin. Note also that StatMech will adjust the value of U to be a value close to what you specified but that also can be divided into an integer number of equally sized bins.

This division into "bins" not only keeps table sizes tractable but also reflects practical realities. As U becomes larger, one will eventually reach a point where distinguishing whether U_A is 500000ε or 500001ε becomes impractical. Putting macropartitions into bins means grouping them into sets whose energies differ by credibly measurable amounts.

Even so, by playing with the controls, you will see that you can make the bin sizes smaller to get a closer look at the "central maximum" (the most probable macropartitions) or the macropartitions at the table's beginning or the end. You can also select whether N_A and N_B refer to atoms or individual oscillators, or whether the graph displays the probability or the logarithm of the probability as a function of U_A/U.

StatMech's capabilities and layout will almost certainly evolve. See the help page associated with the application for the latest information.

T2.7   The Emergence of Irreversibility

With the help of StatMech, we are finally in a position to understand how irreversibility emerges in the case of two Einstein solids in thermal contact.

The case shown in figure T2.4 involves scaling up N_A, N_B, and U by a factor of 1000 compared to the case shown in table T2.1. When $N_A = 1000$, we are still talking about an incredibly tiny speck of a solid only 10 atoms on a side. However, you may have noted that even for this tiny speck the multiplicity numbers have become *outrageously* large (indeed, so large that StatMech must use specially written mathematical software just to evaluate them!). If we increase the system's size by another factor of 100 (so that each solid now has about 46 atoms on a side), StatMech tells us that the total number of microstates available to the system is about 1.80×10^{361232}. Words struggle to express how *huge* this number is. We would need about 80 single-spaced pages *just to write it down* without scientific notation. The entire visible universe only contains about 10^{80} elementary particles, so we would need 10^{361152} universes to contain as many particles as this incredibly tiny system

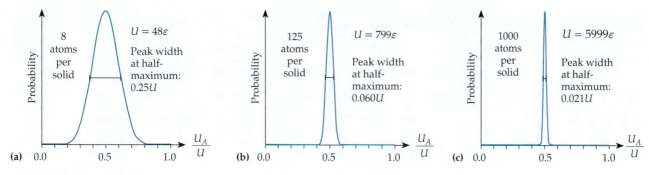

Figure T2.5
How the graph of macropartition probability changes as the system's size increases.

has microstates! Now try to imagine what happens as we increase the system's size by the additional factor of 10^{20} or so we would need to yield even relatively small solids on the everyday scale of things.

A graph of macropartition probabilities as a function of U_A/U also changes shape as we increase the system's size. Figure T2.5 shows a sequence of graphs (traced from StatMech graphs) showing how the probabilities of different macropartitions vary with U_A/U for solids whose number of atoms N_A and N_B and total energy U has been increased by a factor of 8, 125, and 1000 compared to the case shown in table T2.1 (these sizes correspond to each solid being a cube 2, 5, and 10 atoms on a side, respectively). You can see that as the system's size increases, the probability curve becomes an increasingly narrow bell curve centered on the most probable macropartition. You can perhaps imagine that increasing the system's size by another factor of 10^{20} or so will reduce the graph to being a spike of nearly negligible width.

These results have two direct and very important implications:

1. If the combined system is *not* near the most probable macropartition initially, it will rapidly and inevitably move toward that macropartition.
2. It will subsequently *stay* very near to that macropartition in spite of the random shuffling of energy between the two solids.

How do these statements follow from what we have observed?

Let's consider the second implication first. Suppose that our two Einstein solids have nearly the same energy before we bring them into contact. After we do, they randomly shuffle energy around internally and across the boundary between them, sampling various possible microstates. The fundamental assumption implies that the probability that the system will end up in a given microstate after one of these shuffles is the same for all microstates, and since some microstates are in different macropartitions, the combined system's macropartition will **fluctuate** randomly in time.

In the very small system that figure T2.3 describes ($N_A = N_B = 1$, $U = 6\varepsilon$), these fluctuations might be quite large. If we peek at the system at some later time, we will see the system to have randomly wandered into even one of the extreme macropartitions more than 12% of the time. But if we were to peek at the system corresponding to figures T2.4 and T2.5b ($N_A = N_B = 1000$, $U = 5999\varepsilon$), we would find that, more than 99.9% of the time, the system is in a macropartition within the range $U_A/U = 0.500 \pm 0.025$. The probability that we will catch the system in any macropartition where $U_A > 0.60U$ or $U_A < 0.4U$ is smaller than about 10^{-27}. This may seem possible, but it is *not* in fact possible in any realistic sense. If we were to check the system a million times a second for the age of the universe ($\sim 10^{18}$ s), we would *still* have only a 1/1000 chance of seeing the system in such an extreme macropartition.

A system in equilibrium stays in equilibrium

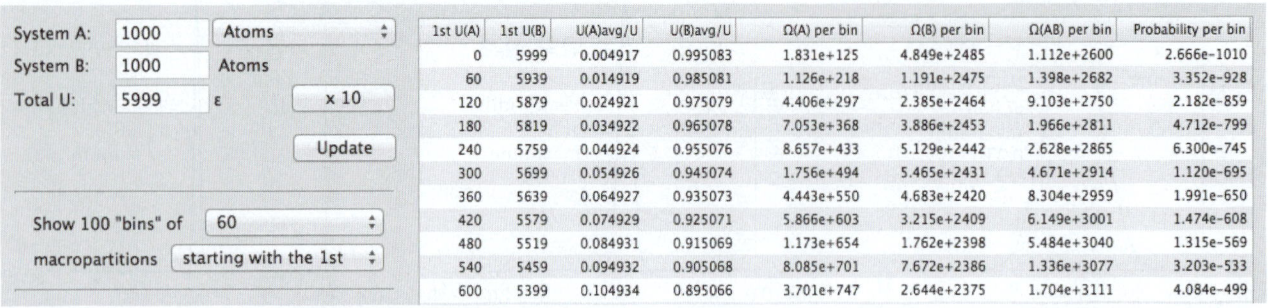

System A:	1000	Atoms
System B:	1000	Atoms
Total U:	5999	ε　　　× 10

Update

Show 100 "bins" of 60

macropartitions　starting with the 1st

1st U(A)	1st U(B)	U(A)avg/U	U(B)avg/U	Ω(A) per bin	Ω(B) per bin	Ω(AB) per bin	Probability per bin
0	5999	0.004917	0.995083	1.831e+125	4.849e+2485	1.112e+2600	2.666e-1010
60	5939	0.014919	0.985081	1.126e+218	1.191e+2475	1.398e+2682	3.352e-928
120	5879	0.024921	0.975079	4.406e+297	2.385e+2464	9.103e+2750	2.182e-859
180	5819	0.034922	0.965078	7.053e+368	3.886e+2453	1.966e+2811	4.712e-799
240	5759	0.044924	0.955076	8.657e+433	5.129e+2442	2.628e+2865	6.300e-745
300	5699	0.054926	0.945074	1.756e+494	5.465e+2431	4.671e+2914	1.120e-695
360	5639	0.064927	0.935073	4.443e+550	4.683e+2420	8.304e+2959	1.991e-650
420	5579	0.074929	0.925071	5.866e+603	3.215e+2409	6.149e+3001	1.474e-608
480	5519	0.084931	0.915069	1.173e+654	1.762e+2398	5.484e+3040	1.315e-569
540	5459	0.094932	0.905068	8.085e+701	7.672e+2386	1.336e+3077	3.203e-533
600	5399	0.104934	0.895066	3.701e+747	2.644e+2375	1.704e+3111	4.084e-499

Figure T2.6
A reprise of figure T2.4.

(The latter is very roughly the probability that you will need emergency treatment during the coming year because you were injured by a pillow or bed mattress.) The numbers become only more extreme as the systems get larger.

So, we see that if two macroscopic solids are initially at or very near the most probable macropartition, they will not stray very far from that macropartition. This is because the *vast majority* of system's microstates are in macropartitions close to the most probable one.

Why the system will move inexorably toward equilibrium

Now suppose that we start with the solids in a fairly extreme macropartition. Once we bring the solids into thermal contact, they will begin to randomly shuffle energy back and forth, sampling new microstates. But the vast majority of microstates near the initial macropartition are to be found in the direction toward the most probable macropartition, so the system will with virtual certainty move in that direction.

For example, consider again the system whose macropartition table appears in figure T2.4 (which I have repeated as figure T2.6 above for convenience). Suppose that, initially, solid A has $U_A \approx 0.015U$, somewhere in the macropartition "bin" corresponding to the table's second row. If random fluctuations in the energy flow between solids A and B cause the system to move to a macropartition in an adjacent bin, what is the probability that it moves to the first row (farther from the probability peak) compared to going to the third row (closer to the probability peak)? If all microstates in these bins are equally probable, then the probability is

$$\frac{\text{Pr(Bin 1)}}{\text{Pr(Bin 3)}} = \frac{1.112 \times 10^{2600}}{9.103 \times 10^{2750}} = 0.1222 \times 10^{-150} = 1.222 \times 10^{-151} \qquad \text{(T2.10)}$$

(Since the exponents are too large for my calculator, I first divided 1.112 by 9.103 to get 0.1222, and then subtracted the exponents to get 10^{-150}.) This probability is so small that for all practical purposes, it is zero. If you checked each of a billion copies of the system a billion times per second for as long as the universe has existed, your probability of seeing this happen would still be only 10^{-115}. Even this latter probability is roughly a millionth of the probability of your winning the California Lottery jackpot 13 times in a row. This is pretty much the operational definition of "never." You will *never ever* see even this tiny, tiny system (a mere 10 atoms on a side) move as much as a hundredth part of U further from equilibrium as opposed to closer to equilibrium! As usual, the numbers become only more outrageously extreme if we scale the system's size upward by another factor of 10^{20}.

How irreversibility emerges from random microscopic energy exchanges

So even random energy exchanges will cause a system of two identical macroscopic Einstein solids to march inexorably and irreversibly toward an equilibrium where both solids' energies are the same, and then never significantly depart from that equilibrium, because *that is where (virtually) all the microstates are*. The huge *differences* in the multiplicities of different macrostates are the ultimate cause of irreversibility in our paradigmatic thermal process.

TWO-MINUTE PROBLEMS

T2T.1 Consider a system consisting of two Einstein solids P and Q in thermal contact. Assume that we know the number of atoms in each solid and ε. What do we know about the system if we also know the quantum state of each atom in each solid?
A. Its macrostate
B. Its microstate
C. Its macropartition
D. Its microstate and macropartition
E. Its macrostate and macropartition
F. Its macrostate and microstate
T. Its macrostate, microstate, and macropartition

T2T.2 Consider a system consisting of two Einstein solids P and Q in thermal contact. Assume that we know the number of atoms in each solid and ε. What do we know about the system if we also know the total energy in each of the two objects? (Please choose from the possible answers listed in problem T2T.1.)

T2T.3 Consider a system consisting of two Einstein solids P and Q in thermal contact. Assume that we know the number of atoms in each solid and ε. What do we know about the system if we also know the total energy of the combined system? (Please choose from the possible answers listed in problem T2T.1.)

T2T.4 What is the *crucial* characteristic of an Einstein solid that makes it easier to analyze in the context of this chapter than most other kinds of thermodynamic systems?
A. The atoms are arranged in a regular, cubic lattice.
B. The atoms are identical.
C. Its microstates are comparatively easy to count.
D. Each atom's energy levels are equally spaced.
E. $\Omega(U, N)$ is always reasonably small.
F. Other (specify).

T2T.5 We can ignore an Einstein solid's zero-point energy because
A. it is zero.
B. it never changes in any thermal interaction.
C. it is insignificant compared to the solid's total energy.
D. it is just a quantum-mechanical effect.
E. other (specify).

T2T.6 Which of the following statements is true?
A. There are always many microstates in a macrostate.
B. All accessible macrostates are equally probable.
C. All microstates of a system are equally probable.
D. All accessible macropartitions are equally probable.
E. When two objects in thermal contact are isolated from everything else, their macrostates cannot change.
F. None of the above.

T2T.7 Suppose that we increase an N-atom Einstein solid's value of $q = U/\varepsilon$ from q to $q + 1$. By what factor does the value of its multiplicity Ω increase? (Choose whichever result is the closest, assuming that $q \gg 1$ and $N \gg 1$.)
A. q
B. $3N$
C. $q/3N$
D. $3N/q$
E. $(q + 3N)/q$
F. $(q + 3N)/(qN)$

T2T.8 In the situation shown in figure T2.5c, the width of the bell curve at half its peak value is $0.021U$. If we multiply N_A, N_B, and U by a factor of 100, then the peak's width (as you can check) is about $0.0021U$. Suppose that we increase each solid's N and the system's total energy by *another* factor of 10^{18} to create solids containing $1/6$ of a mole of atoms (still pretty small objects by everyday standards). Assuming that the trend continues, what will be the approximate width of the combined system's probability bell curve as a fraction of U?
A. 2×10^{-20}
B. 2×10^{-22}
C. 2×10^{-10}
D. 2×10^{-12}
E. 2×10^{-40}
F. 2×10^{-42}
T. Some other factor (specify).

T2T.9 Suppose that you use a super-fast computer to measure a system's macropartition a billion times a second. What is the approximate probability of the least-probable macropartition that you might plausibly see in your lifetime? Select the closest response.
A. 1 chance in a million (10^6)
B. 1 chance in a billion (10^9)
C. 1 chance in a trillion (10^{12})
D. 1 chance in 10^{15}
E. 1 chance in 10^{18}
F. 1 chance in 10^{21}

T2T.10 Consider the system where $N_A = N_B = 1000$, $U = 5999\varepsilon$ (see figure T2.6). Suppose that we can measure each object's energy to within $0.01U \approx 60\varepsilon$ (the size of one bin). Once the system has reached a macropartition in the most probable bin, it will never spontaneously move to a macropartition outside that bin. T or F? (*Hint:* Scroll down the table in StatMech to find the most probable bin, and look at the probabilities of adjacent bins.)

T2T.11 Which of the systems listed below is the largest, having $N_A = N_B$ and $U = 6N_A\varepsilon$ for which there is a better than 1 in a billion chance of seeing one or the other solid having zero energy? (*Hint:* Using StatMech to check cases is probably the easiest approach.)
A. $N_A = 6$
B. $N_A = 8$
C. $N_A = 10$
D. $N_A = 12$
E. $N_A = 15$

HOMEWORK PROBLEMS

Basic Skills

T2B.1   Calculate the multiplicity of an Einstein solid with $N = 1$ and $U = 6\varepsilon$ by directly listing and counting the microstates. Check your work by using equation T2.7.

T2B.2   Calculate the multiplicity of an Einstein solid with $N = 1$ and $U = 5\varepsilon$ by directly listing and counting the microstates. Check your work by using equation T2.7.

T2B.3   Use equation T2.7 to calculate the multiplicity of an Einstein solid with $N = 4$ and $U = 10\varepsilon$.

T2B.4   Use equation T2.7 to calculate the multiplicity of an Einstein solid with $N = 3$ and $U = 15\varepsilon$.

T2B.5   If the fundamental assumption of statistical mechanics is true, how many times more likely is the combined system of solids described in table T2.1 to be found in the macropartition where $U_A = U_B = 3\varepsilon$ than in the macropartition where $U_A = 0$, $U_B = 6\varepsilon$?

T2B.6   Consider the system consisting of a pair of Einstein solids in thermal contact. A certain specific macropartition has a multiplicity of 3.7×10^{1024}, while the total number of microstates available to the system in *all* macropartitions is 5.9×10^{1042}. If we look at the system at a given instant of time, what is the probability that we will find it to be in our specific macropartition?

T2B.7   Consider the system consisting of a pair of Einstein solids in thermal contact. A certain macropartition has a multiplicity of 1.2×10^{346}, while the total number of microstates available to the system in all macropartitions is 5.9×10^{362}. If we look at the system at a given instant of time, what is the probability that we will find it to be in our certain macropartition?

T2B.8   Consider the system consisting of a pair of Einstein solids in thermal contact. Suppose that the system is initially in a macropartition whose multiplicity is 8.8×10^{123}. The adjacent macropartition closer to the equilibrium macropartition has a multiplicity of 4.2×10^{134}. If we look at the system a short time later, how many times more likely is it to have moved to the second macropartition than to have stayed with the first?

T2B.9   Consider the system consisting of a pair of Einstein solids in thermal contact. Suppose that it is initially in a macropartition whose multiplicity is 7.6×10^{3235}. The adjacent macropartition closer to the equilibrium macropartition has a multiplicity of 4.1×10^{3278}. If we look at the system a short time later, how many times more likely is it to have moved to the second macropartition than to have stayed with the first?

Modeling

T2M.1   Imagine putting the two solids discussed in problems T2D.1 and T2D.2 into thermal contact. Suppose that the resulting combined system is isolated from everything else, and that the combined system contains 6 units of energy (that is, $U_A + U_B = 6\varepsilon$).
(a) Construct a table showing U_A, U_B, Ω_A, Ω_B, and Ω_{AB} for all possible macropartitions of the system (that is, a table analogous to table T2.1), and compute the probabilities for each of the seven possible macropartitions according to the fundamental assumption.
(b) Identify the most probable macropartition(s) of this system. Is the energy evenly divided between the solids' atoms in the most probable macropartition(s)?

T2M.2   Imagine putting the two solids discussed in problems T2D.1 and T2D.2 into thermal contact. Imagine that the resulting combined system is isolated from everything else, and that the combined system contains 9 units of energy (that is, $U_A + U_B = 9\varepsilon$).
(a) Construct a table showing U_A, U_B, Ω_A, Ω_B, and Ω_{AB} for all possible macropartitions of the system (that is, a table analogous to table T2.1), and compute the probabilities for each of the 10 possible macropartitions according to the fundamental assumption.
(b) Identify the most probable macropartition(s) of this system. Is the energy evenly divided between the solids' atoms in the most probable macropartition(s)?

T2M.3   Imagine putting two solids with $N_A = N_B = 2$ in thermal contact. Suppose that the resulting combined system is isolated from everything else and that it contains a total of 9 units of energy (that is, $U_A + U_B = 9\varepsilon$).
(a) Construct a table showing U_A, U_B, Ω_A, Ω_B, and Ω_{AB} for all possible macropartitions of the system (that is, a table analogous to table T2.1), and compute the probabilities for each of the 10 possible macropartitions according to the fundamental assumption. You may use the results of problem T2D.2.
(b) Identify the most probable macropartition(s) of this system. Is the energy evenly divided between the solids' atoms in the most probable macropartition(s)?

T2M.4   How does size matter? For each of the Einstein solid pairs described in parts (a) through (c), use StatMech to answer the following questions:
(1) How many total microstates are available to the combined system?
(2) Which are the two most probable macropartition bins, and how many total microstates are available to the system in these two bins?
(3) What is the approximate average energy per atom in each solid if the system's macropartition is in one of these most probable bins?

(4) What is the width (as a fraction of U) of the probability curve's peak at half the peak's maximum value? Answer these four questions for the pair

(a) where $N_A = N_B = 20$, $U = 99\varepsilon$,

(b) where $N_A = N_B = 2000$, $U = 9999\varepsilon$,

(c) and where $N_A = N_B = 200{,}000$ and $U = 999{,}999\varepsilon$.

(d) Note that each case is 100 times larger in both size and energy (approximately) than the previous case. Carefully and as quantitatively as you can describe any trends you see in how the answers to the four questions above change as the system becomes larger.

T2M.5 Does relative size matter?

(a) Use StatMech to generate tables for the following five different Einstein solid pairs.

(1) $N_A = N_B = 50$

(2) $N_A = 60$, $N_B = 40$

(3) $N_A = 70$, $N_B = 30$

(4) $N_A = 80$, $N_B = 20$

(5) $N_A = 90$, $N_B = 10$

In each case, choose $U = 100\varepsilon$. Compare the average energy per atom (that is, compare U_A/N_A to U_B/N_B) for the value of U_A corresponding to the peak probability as reported by StatMech below the graph. (Note that StatMech actually interpolates where the peak *would* be if the probability function were continuous and smooth, so U_A might not be an integer.)

(b) State a simple approximate rule describing how U_A/N_A compares to U_B/N_B in the most probable macropartition. Is your rule always *exactly* true?

(c) How does your rule fare when you increase all numbers (N_A, N_B, and U) simultaneously by a factor of 100? Does its accuracy improve for the more extreme cases?

(d) How do you think your rule will fare in the limit that $N_A + N_B$ becomes extremely large (as would be the case for actual macroscopic solids)?

T2M.6 In this problem, we will use StatMech to examine the effects of *separately* varying an Einstein solid system's size and total energy in cases where the system has either lots of energy per atom or little energy per atom.

(a) Consider first an energy-poor system of Einstein solids in contact where $N_A = N_B = 1000$, and $U = 100\varepsilon$. Which has the greater affect on (1) the probability peak's width at half maximum (as a fraction of U) and (2) the probability of the macropartition row or bin where U_A is smallest: multiplying both N_A and N_B by 100 or multiplying the energy by 100? Support your answer by describing what happens in each case.

(b) Now consider an energy-rich system where $N_A = N_B = 100$ and $U = 999\varepsilon$. Now which has the greater effect?

(c) *Explain* what you observe (as well as you can) in plain language: *Why* do you think that the system behaves this way? (Don't worry too much about getting your explanation "right:" the goal is simply for you to *try* to work toward an intuitive understanding.)

T2M.7 The width of the probability curve at half its maximum value for a pair of identical Einstein solids as a fraction of U turns out to be of the order of magnitude of

$\sqrt{1/N}$, where N is the number of atoms in each solid. Consider solids with $N \approx 10^{22}$ atoms, which is still a pretty small solid on the macroscopic scale ($1/60$ of a mole!).

(a) What will be the approximate width (as a fraction of U) of the probability curve for a pair of such solids in thermal contact?

(b) Suppose that we wanted to draw a graph where the peak was 1 mm wide. How long would we have to make the horizontal axis (assuming that the axis stretches from 0 to U)?

(c) Suppose we want to fit the graph on a page. Roughly how wide would the peak be?

T2M.8 The probability of drawing a royal flush in poker is 1.54×10^{-6}. Now, consider a macropartition with probability 10^{-93}. If this probability is equal to drawing n royal flushes in a row, what is n?

T2M.9 Suppose that the probability of a certain macropartition is 2.29×10^{-100}. If this is equivalent to the probability of throwing a fair coin and coming up heads n times in a row, what is n?

T2M.10 You have a computerized measuring device that can measure and record the macropartition of an Einstein solid system a billion times a second. You run the experiment for 30 years. What is the order of magnitude of the probability of the least-probable macropartition that you might plausibly see in your data set?

Derivation

T2D.1 Consider an Einstein solid consisting of $N_A = 1$ atom (three oscillators). Each oscillator can store any integer number of energy units ε. The following table lists the number of microstates Ω_A available to the solid when it has various values of total thermal energy U_A.

U_A	0	1ε	2ε	3ε	4ε	5ε	6ε	7ε	8ε	9ε
Ω_A	1	3	6	10	15	21	28	36	45	55

By actually listing and counting the various possible microstates, verify the results for the multiplicity Ω_A for the cases where $U_A = 4\varepsilon$ and $U_A = 7\varepsilon$. (You can check your results by using equation T2.7.)

T2D.2 Consider an Einstein solid consisting of $N_B = 2$ atoms. The following table lists the number of microstates Ω_A available to the solid when it has various values of total internal energy U_B.

U_A	0	1ε	2ε	3ε	4ε	5ε	6ε	7ε	8ε	9ε
Ω_A	1	6	21	56	126	252	462	792	1287	2002

(a) By actually listing the various possible microstates, verify the result for $U_B = 2\varepsilon$.

(b) Using equation T2.7, verify the value of Ω_B for $U_B = 6\varepsilon$ and 9ε.

T2D.3 We can derive equation T2.7 as follows. First, note that the problem of counting the microstates of an Einstein solid is essentially the same as the problem of finding the number of distinct patterns that can be generated by pulling marbles and matchsticks randomly from a bag. For example, suppose that we pull the following sequence of items from the bag (reading from left to right):

If we imagine each marble to be a unit of energy and each matchstick to be a *division* between two oscillators, then this pattern corresponds to the 130211 microstate for an Einstein solid with $N = 2$ atoms (six oscillators) and $U = 8\varepsilon$.

(a) Argue that a solid with M oscillators and q units of energy will be represented in this scheme by q marbles and $M - 1$ matches.

(b) Suppose that we put $M - 1$ matches and q marbles in the bag. Argue that we can pull objects out of the bag a total of $(M + q - 1)!$ different ways. (*Hint:* When we select the first object, we have $M + q - 1$ choices. When we choose the second item, we now only have $M + q - 2$ choices, since we've already pulled out one object.)

(c) Not all these distinct ways of pulling out objects generate distinct patterns. For example, consider taking a given pattern and rearranging the marbles. The rearrangement does not change the basic pattern, but would represent a different sequence of choices as we pull objects out of the bag, and thus would be counted as a distinct choice in (b). Argue that there are $q!$ ways of rearranging the marbles and $(M - 1)!$ ways of rearranging the matchsticks without affecting the pattern.

(d) Argue finally that equation T2.7 correctly states the number of distinct patterns that can be generated, and thus the number of distinct microstates of an Einstein solid with M oscillators and n units of energy.

T2D.4 Consider equation T2.7.

(a) Argue that if you increase the value of q by 1, then the value of the multiplicity Ω will increase by a factor of $(q + 3N)/(q + 1)$.

(b) By what factor does Ω increase if you increase the value of N by 1?

T2D.5 (a) Show that taking the time derivative of equation T2.1 yields

$$0 = v_x(ma_x + k_s x) + v_y(ma_y + k_s y) + v_z(ma_z + k_s z) \quad \text{(T2.11)}$$

(b) By changing initial conditions, I could arrange it so that v_x, v_y, and v_z have any values I please at a given time. Argue that since this equation has to be zero at *all* times, the quantities in parentheses have to be *independently* equal to zero at all times.

(c) Show that each of these quantities in parentheses is the same as Newton's second law for a simple one-dimensional harmonic oscillator moving in the corresponding axis direction.

T2D.6 We can estimate the approximate size of a macroscopic system's fluctuations around the equilibrium macropartition as follows. When an object A has a very large number of molecules, it tends (no matter what substance it is made of) to have a multiplicity of the form $\Omega_A \approx C U_A^{bN}$, where U_A is the object's thermal energy, N is the number of molecules it contains, C is a quantity that depends on the type of substance and on N (and maybe other macroscopic variables) but not on U_A, and b is a substance-dependent constant of the order of magnitude of 1. Consider two identical objects (A and B) in thermal contact.

(a) Argue that we can write the multiplicity of a combined system with total energy U as

$$\Omega_{AB} \approx D(1 - x^2)^{bN} \qquad \text{where } x \equiv \frac{U_A - \frac{1}{2}U}{\frac{1}{2}U} \quad \text{(T2.12)}$$

and D is a quantity that depends on U, but not on x, and N is the number of molecules in each solid. Note that the variable x here specifies how far we are from equilibrium. If $x = 0$, then $U_A = \frac{1}{2}U$, which is the equilibrium macropartition, and $x = \pm 1$ corresponds to the extreme macropartitions $U_A = U$ and $U_A = 0$, respectively. Note that Ω_{AB} has a peak at $x = 0$ and falls off very rapidly as the absolute value of x grows.

(b) Show the value of x where Ω_{AB} has fallen to one-half of its peak value is

$$x_{1/2} = \pm \frac{1}{\sqrt{2bN}} \quad \text{(T2.13)}$$

(*Hint:* The famous "binomial approximation" states that $(1 - z)^n \approx 1 - nz$ as long as $z \ll 1$. Argue that one can use this approximation to calculate the binomial here.)

(c) Assuming that $b \approx 1$ and $N \approx 10^{24}$, estimate very roughly how far a typical fluctuation will carry the energy of either system away from its equilibrium value $\frac{1}{2}U$, expressed as a fraction of that equilibrium value.

Rich-Context

T2R.1 Consider a hypothetical system in which each atom can only be in one of two quantum states, a ground state with energy 0 or an excited state with energy ε. The atoms have fixed positions and fixed volumes, and there is no way for an atom to change its energy except by going from the ground state to the excited state or vice versa. As with the Einstein solid, we can completely describe the macrostate of this system by specifying the number of atoms N and their total thermal energy U.

(a) Find an expression, in terms of U and ε, for the number of atoms n that are in the excited state.

(b) Find a formula for the multiplicity $\Omega(U, N)$ of a macrostate where the system has N atoms and total energy U. (*Hints:* Play around with small systems first. For example, suppose we have $N = 3$ and $U = 2\varepsilon$. We might describe the possible microstates as being 011, 101, and 110. Once you have developed some sense about the multiplicity of small systems, start working on a general formula, which will involve some factorials. Once you have a trial formula, go back and check it in small-system cases where you can directly count the microstates.)

(c) Construct a macropartition table for a system consisting of two subsystems with $N_A = N_B = 20$ and $U = 8\varepsilon$. Does it look at least qualitatively like the macropartition tables we have been constructing for Einstein solids? Explain.

Comment: This model is also actually a pretty good description of certain kinds of solids at very low temperatures when they are placed in a strong magnetic field. Under such circumstances, the magnetic moment of an atom in the solid can be either aligned with the field (which we can take to be the ground state) or anti-aligned with the field (which we can take to be the excited state). It is also easier to determine multiplicities for this model than for the Einstein solid. So why do we not focus on this model instead of on the Einstein solid? This model only applies to pretty esoteric systems. Even worse, it exhibits peculiar behavior when its energy is such that $n > \frac{1}{2}N$, behavior that is almost never seen in nature otherwise. Therefore, unlike the Einstein solid, this model is not very useful for helping us understand how normal complex systems behave.

T2R.2 (Adapted from Kittel and Kroemer, *Thermal Physics*, 2d ed. Macmillan, 1980, p. 53.) The writer Aldous Huxley is reported to have said that "six monkeys, set to strum unintelligently on typewriters for millions of years would be bound in time to write all of the books in the British Museum." This statement is in fact completely *false*: Huxley has been misled by an incorrect intuition about the character of extremely large numbers.

Let us set ourselves a much less challenging task. Suppose we have 10 billion monkeys (somewhat more than the human population of the earth) typing diligently at the rate of 2 characters per second since the universe began 5×10^{17} s ago. Instead of requiring an entire library, we will settle for a single typed version of *Hamlet*. Let us guess that *Hamlet* has approximately 10^5 characters.

(a) Assume that the typewriters used by the monkeys have 26 letters and 10 punctuation characters (space, carriage return, period, comma, colon, semicolon, quotation mark, apostrophe, dash, question mark) for a total of 36 characters. We will ignore the distinction between capital letters and small letters. The probability that any given character is the first character in *Hamlet* is thus 1/36. The probability that this character and the next are the first *two* characters of *Hamlet* is $(1/36)(1/36) = (1/36)^2$. Argue, then, that the probability that any given sequence of 10^5 random characters happens to be *Hamlet* is $10^{-155,630}$. (*Hint:* $\log x^a = a \log x$, where log is the base-10 logarithm.)

(b) Suppose that we paste all the pages generated by the monkeys in a single sequence so that *any* key typed by *any* monkey could in principle be the first character in the play. Argue that the probability that *Hamlet* would appear *somewhere* in our sequence of pages is $10^{-155,602}$. This probability is *zero* in any practical sense: Even our huge bevy of monkeys will never, never, *never* be able to type *Hamlet* at random.

ANSWERS TO EXERCISES

T2X.1 111, 012, 021, 102, 201, 120, 210, 003, 030, and 300, so $\Omega(1, 3\varepsilon) = 10$

T2X.2 $\Omega(3, 8\varepsilon) = 12{,}870$

T2X.3 The table is as follows:

U_A	U_B	Ω_A	Ω_B	Ω_{AB}
0	8	1	45	45
1	7	3	36	108
2	6	6	28	168
3	5	10	21	210
4	4	15	15	225
5	3	21	10	210
6	2	28	6	168
7	1	36	3	108
8	0	45	1	45

Total number of microstates = 1287

T3

Entropy and Temperature

Chapter Overview

Section T3.1: The Definition of Entropy

We define a system's **entropy** S in a given macrostate to be

$$S \text{ (of a macrostate)} \equiv k_B \ln \Omega \text{ (of the macrostate)} \qquad \text{(T3.1)}$$

- **Purpose:** This equation defines the entropy S of a thermodynamic system in a given macrostate, where Ω is the macrostate's multiplicity; k_B is Boltzmann's constant: $k_B = 1.38 \times 10^{-23}$ J/K.
- **Limitations:** Since this is a definition, it has no limitations. However, determining Ω exactly is difficult in many situations.

Entropy S is just a different way to describe a system's multiplicity Ω, but it has several advantages over working with the multiplicity directly: (1) the logarithm converts extremely large multiplicities to more manageably sized entropies; and (2) when we put two objects in contact, the total system's multiplicity is the *product* of the two objects' multiplicities, but the total entropy is the *sum* of the individual entropies:

$$S_{AB} = S_A + S_B \qquad \text{(T3.3)}$$

- **Purpose:** This equation describes the total entropy S_{AB} for a given macropartition of a system consisting of two subsystems A and B in thermal contact, where S_A and S_B are those subsystems' entropies in that macropartition.
- **Limitations:** There are none: this directly follows from the definitions of entropy and multiplicity.

The section also distinguishes between *ordinary numbers*, *large numbers* (to which one can add an ordinary number without significantly affecting it) and *ginormous* numbers (which one can multiply even by a large number without significantly affecting it). Macroscopic objects have ginormous multiplicities, but ordinary entropies.

Section T3.2: The Second Law of Thermodynamics

We have seen that random energy transfers between large systems in thermal contact essentially *never* move the combined system to a macropartition with a significantly smaller multiplicity (and thus smaller total entropy). The **second law of thermodynamics** expresses this truth in a short, simple phrase: *The entropy of an isolated system never decreases.* This is one of the most important and useful laws in physics. However, one should always remember that this law (as we have seen) is a consequence of more fundamental principles (that all microstates are equally probable and that the multiplicities of most macroscopic objects increase incredibly rapidly with increasing energy).

Section T3.3: Entropy and Disorder

Equation T3.3 links entropy to *multiplicity*. Popular treatments of entropy often link it to *disorder*. This often works because macrostates we consider to be "disordered" *usually* have greater multiplicities than states we consider to be "ordered." But this linkage can also be misleading: sometimes macrostates that *appear* more orderly than others actually have more microstates. When in doubt, you should always remember that entropy is most fundamentally linked to *multiplicity*, not disorder.

Section T3.4: The Definition of Temperature

The total entropy $S_{\text{TOT}} = S_A + S_B$ of an isolated system consisting of two objects A and B is a maximum (by definition!) when the system is in its equilibrium (most probable) macropartition. Therefore, in equilibrium we must have

$$0 = \frac{dS_{\text{TOT}}}{dU_A} = \frac{dS_A}{dU_A} + \frac{dS_B}{dU_A} = \frac{dS_A}{dU_A} - \frac{dS_B}{dU_B} \quad \Rightarrow \quad \frac{dS_A}{dU_A} = \frac{dS_B}{dU_B} \qquad \text{(T3.9)}$$

(The next-to-last step follows because $dU_A = -dU_B$ in this isolated system.) By the zeroth law, the two objects' temperatures (by definition!) must be equal in that macropartition as well. So an object's value of dS/dU must be some function $f(T)$ of that object's temperature T. The choice $f(T) \equiv 1/T$ is the simplest one consistent with the idea that heat spontaneously flows from an object with a large value of T to an object with a lower value of T. So (with Kelvin) we *define* an object's temperature as follows:

$$\frac{1}{T} \equiv \frac{dS}{dU} \text{ (with other variables constant)} \equiv \frac{\partial S}{\partial U} \qquad \text{(T3.12)}$$

- **Purpose:** This equation defines an object's absolute temperature T in terms of the rate at which the object's entropy S increases with its internal energy U.
- **Limitations:** We must hold the object's other macroscopic variables (the number of molecules N, volume V, and so on) constant when evaluating the derivative (this is what the partial derivative notation $\partial S/\partial U$ means).

Because in principle we can *calculate* an object's multiplicity and thus its entropy, we can calculate an object's temperature T without referring to a standard thermoscope.

Section T3.5: Consistency with Historical Definitions

A calculation based on equation T3.12 shows that $dU/dT \approx 3Nk_B$ for an Einstein solid at sufficiently high temperatures. This is consistent with historical thermoscope results discussed in chapter T1, suggesting that the new definition of temperature is at least approximately the same as historical thermosope-based definitions.

Section T3.6: A Financial Analogy

Objects in thermal contact behave like people exchanging money do if

Energy	$\leftrightarrow$	Money
Entropy	$\leftrightarrow$	Happiness
Temperature	$\leftrightarrow$	Generosity

and we assume that money gets distributed in a way that increases the total happiness of the community. Just as total happiness increases if money flows from a more generous person to a more needy person, a thermodynamic system's entropy increases if energy flows from a high-temperature object to a lower-temperature one.

Most normal objects have entropies and temperatures that increase with energy, just as normal people's happiness and generosity increase when they receive money, but physical systems that are "miserly" (whose temperature decreases as their energy increases) or "enlightened" (whose entropy increases as their energy decreases) do exist. The analogy can help us predict about how such exotic objects behave.

T3.1   The Definition of Entropy

In the last chapter, we saw why the paradigmatic process of thermal physics (two objects coming into equilibrium) is irreversible: the system's macropartitions near the most probable macropartition (the equilibrium energy distribution) have *vastly* more microstates (and thus vastly greater probability) than macropartitions far from equilibrium. The enormous disparity between the numbers of microstates available in various macropartitions makes the march toward the equilibrium macrostate completely inevitable.

However, the same incredible immensity of macrostates' multiplicities also makes these multiplicities awkward to deal with. Indeed, the number of microstates available to Einstein solids involving Avogadro's number of atoms are in the ballpark of $10^{10^{23}}$ (that is, a 1 followed by Avogadro's number of zeros). Such numbers are truly unmanageable. For realistically sized objects, we need a less awkward way to talk about multiplicity.

To this end, we define an object's **entropy** S in a given macrostate to be:

The definition of entropy

$$S \text{ (of a macrostate)} = k_B \ln \Omega \text{ (of the macrostate)} \qquad (T3.1)$$

- **Purpose:** This equation defines the entropy S of a thermodynamic system in a given macrostate, where Ω is the macrostate's multiplicity; k_B is Boltzmann's constant: $k_B = 1.38 \times 10^{-23}$ J/K.
- **Limitations:** Since this is a definition, it has no limitations. However, determining Ω exactly is difficult in many situations.

In the specific case of an Einstein solid, whose macrostate is specified by U and N, its entropy is a function of U and N:

Entropy of an Einstein solid

$$S(U, N) \equiv k_B \ln \Omega(U, N) = k_B \ln \frac{(3N + U/\varepsilon - 1)!}{(3N - 1)!(U/\varepsilon)!} \qquad (T3.2)$$

where ε is the difference between adjacent energy levels in the solid's oscillators.

Some nice features of this definition

Note that *entropy* is really just another way of talking about *multiplicity*: when the multiplicity is large, the entropy is large. But defining the entropy in terms of the *logarithm* of the multiplicity serves two useful purposes. First, it makes awkwardly large multiplicity values easier to manage. For example, the natural logarithm of a multiplicity $\Omega = 10^{10^{23}}$ is about 2.3×10^{23}. While still a large number, this is much more manageable than the multiplicity itself.

Second, consider a specific macropartition of a system of two systems in thermal contact. The multiplicity of the combined system in this macropartition is $\Omega_{AB} = \Omega_A \Omega_B$, where Ω_A and Ω_B are the multiplicities of the individual systems in their specified macrostates. This implies that the total entropy of the combined system in that macropartition is simply

$$S_{AB} = S_A + S_B \qquad (T3.3)$$

- **Purpose:** This equation describes the total entropy S_{AB} for a given macropartition of a system consisting of two subsystems A and B in thermal contact, where S_A and S_B are those subsystems' entropies in that macropartition.
- **Limitations:** There are none: this directly follows from the definitions of entropy and multiplicity.

Equation T3.3 follows from the definitions of entropy and multiplicity because for any x and y, $\ln xy = \ln x + \ln y$. This means that a macropartition's entropy must be $S = k_B \ln \Omega_{AB} = k_B \ln(\Omega_A \Omega_B) = k_B \ln \Omega_A + k_B \ln \Omega_B = S_A + S_B$, where S_A and S_B are the subsystem's entropies in this macropartition.

Now, we could have defined $S = \ln \Omega$ without the k_B (and a few textbook authors do this). This shares all the advantages of the conventional definition and in a certain sense more fundamentally captures the essence of entropy. The constant k_B just re-scales $\ln \Omega$ in a way that we will see is helpful when we define temperature later in the chapter. For now, simply note that since $k_B = 1.38 \times 10^{-23}$ J/K, a multiplicity on the order of magnitude of $10^{10^{23}}$, whose logarithm is about 2.3×10^{23}, will have a corresponding entropy $S = k_B \ln \Omega$ of about 4 J/K, a conveniently small number.

Indeed, the numbers that occur in statistical mechanics occur in three distinguishably different sizes.* **Small numbers** are ordinary numbers, such as 3, 47, and 23,438. **Large numbers** are numbers on the scale of Avogadro's number $N_A = 6.023 \times 10^{23}$. The basic distinction between large and small numbers is that you can *add* a small number to a large number without significantly changing it: $6.023 \times 10^{23} + 47 = 6.023 \times 10^{23}$. Large numbers require scientific notation to handle easily. (Note that "astronomical numbers," such as the number of stars in a galaxy, or light-years to the most distant galaxies, are at the *small* end of the large number scale.) **Ginormous numbers** are numbers like $10^{10^{23}}$, numbers so huge we can *multiply* them even by a large number and still have basically the same thing:

$$(10^{10^{23}})(10^{23}) = 10^{(10^{23}+23)} = 10^{10^{23}} \tag{T3.4}$$

Ginormous numbers are tricky to handle even in scientific notation, but taking the logarithm of a ginormous number converts it to just a large number, and then multiplying that number by a number like k_B makes it a small number. This is what we are doing with entropy.

Problem: Suppose that we have two Einstein solids in thermal contact with $N_A = N_B = 250$ and $U = 999\varepsilon$. In the single macropartition where $U_A = 12\varepsilon$ (as you can check with StatMech) $\Omega_A = 7.22 \times 10^{25}$, $\Omega_B = 5.84 \times 10^{513}$, and $\Omega_{AB} = 4.22 \times 10^{539}$. **(a)** Find the entropies of both individual solids and the combined system in this macropartition. **(b)** Compare the entropy of the total system in this macropartition with that of the system in its central individual macropartitions (where $\Omega_{AB} = 3.48 \times 10^{726}$). Express all entropies as a multiple of k_B.

Example T3.1

Solution I can calculate S_A directly: $S_A \equiv k_B \ln \Omega_A = 59.54 k_B$. However, my calculator cannot handle exponents greater than 99, so I cannot calculate either $\ln \Omega_B$ or $\ln \Omega_{AB}$ directly. However, because $\ln x^a = a \ln x$, $\ln 10^{513} = 513 \ln 10 = 513(2.03259)$. Moreover, $\ln xy = \ln x + \ln y$, so

$$S_B = k_B \ln \Omega_B = k_B [\ln 5.84 + 513 \ln 10] = 1182.99 k_B \tag{T3.5}$$

Similarly, $S_{AB} = k_B[\ln 4.22 + 539(2.30259)] = 1242.53 k_B$. (Note that we have $S_{AB} = 1242.53 k_B = 1182.99 k_B + 59.54 k_B = S_A + S_B$, as we must.) The system's entropy in a central macropartition is $S_{AB} = k_B[\ln 3.48 + 726(2.30259)] = 1672.92 k_B$, which is substantially larger (as we would expect).

---

*These categories were invented by Daniel Schroeder (*Thermal Physics*, Addison-Wesley-Longman, 2000, p. 61), though I have renamed the last category.

Example T3.2

Problem: Suppose that one macropartition of a combined system of two Einstein solids has an entropy $S_1 = 432.5k_B$ while another (where the energy is more evenly divided) has an entropy of $S_2 = 546.3k_B$. How many times more probable is the second macropartition than the first?

Solution A given macropartition's probability is proportional to its multiplicity Ω. Now, the natural log and exponential functions are inverses, so $e^{\ln x} = x$. Since we define entropy to be $S = k_B \ln \Omega$, this means that $\Omega = e^{S/k_B}$. Therefore, the ratio of the probabilities in this case is

$$\frac{\Omega_2}{\Omega_1} = \frac{e^{S_2/k_B}}{e^{S_1/k_B}} = \frac{e^{546.3}}{e^{432.5}} = e^{546.3-432.5} = e^{113.8} = 2.6 \times 10^{49} \tag{T3.6}$$

Therefore, we are 2.6×10^{49} times more likely to find the system in the second macropartition than in the first.

T3.2   The Second Law of Thermodynamics

We have seen that a system of two objects will evolve from macropartitions with lower multiplicities (and thus lower entropy) to macropartitions with higher multiplicities (higher entropy). Moreover, once the system reaches the macropartition having the highest multiplicity (highest entropy), it will stay very close to that macropartition. For very tiny systems, these are merely statements of probability, but for anything approaching a normal-size system, the extremely large multiplicities turn probabilities in principle to certainties in practice. Therefore, as a large system inevitably evolves from macropartitions with low multiplicities to those with larger multiplicities, *its entropy inevitably increases.* The entropy of a (sufficiently large) system will *never* be observed to decrease, because the idea that the system would evolve to a macropartition where its entropy is measurably smaller is extraordinarily improbable (see problem T3D.1 for a numerical example). The following simple law, known as the **second law of thermodynamics,** expresses the essence of this idea:

> *An isolated system's entropy never decreases.*

The second law of thermodynamics

This law is one of the most important and useful laws of physics and is (as the writer C. P. Snow once asserted) something *any* well-educated person ought to know. It expresses precisely and quantitatively the basis of irreversibility and has many implications and applications (as we will see!). But we should never lose sight of the fact (as Boltzmann first argued) that this law is a *consequence* of the more fundamental ideas presented in chapter T2.

T3.3   Entropy and Disorder

Many popular treatments link *entropy* to the concept of *disorder.* Here we have defined entropy in terms of *multiplicity.* How are these ideas related?

Examples of disorder linked to multiplicity

Think about it this way. Why does your dorm room get disorderly unless you specifically clean it up? *Because that's where the microstates are.* There are many, many more ways for your room to be disorderly than orderly. Because of this, random things that happen in your room are *far* more likely to contribute to disorder than to greater order. (What is the likelihood that an earthquake will pick up your clothes and hang them back up in your closet?)

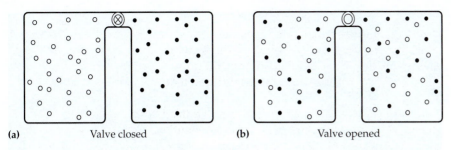

(a) Valve closed   **(b)** Valve opened

Figure T3.1
Spontaneous mixing of gas molecules.

Let us consider some more physical examples of disorder. A flowing liquid of a certain kind of atom has greater disorder than a neat crystalline solid made of the same atoms. A substance with a sufficient amount of thermal energy will therefore be a liquid instead of a solid because *that's where the microstates are:* there are many, many more microstates available to molecules if they are free to roam around than if they are confined to a solid.

Figure T3.1 shows a container with two compartments. Before mixing, each compartment contains different gases. A valve between the two compartments is then opened, allowing the gases to mix with each other. You may know that under such circumstances, the gases will *spontaneously* intermingle, just as a drop of cream put in a cup of coffee will naturally diffuse throughout the cup (although a little stirring speeds up the process). Why do the gases spontaneously mix, becoming more "disorderly" in a certain sense? *Because that's where the microstates are.* There are many, many, many more microstates available to the system when the gas molecules are free to roam over the entire container than there are when the gases are separated.

So in many standard situations, increased multiplicity is indeed linked with disorder. It is important to remember, however, that entropy is defined in terms of *multiplicity*, not disorder. Increasing entropy does not *necessarily* imply increasing disorder (at least *visible* disorder).

For example, consider the glass of crushed ice and the glass of water shown in figure T3.2. Which of the two glasses has a higher entropy? The crushed ice may *look* more disorderly, but since we must add energy to the ice to convert it to water, and since the multiplicity of virtually all macroscopic systems increases dramatically when we add energy, the multiplicity (and thus the entropy) of the nice, orderly-looking glass of water is much larger.

Another potentially confusing example is life itself. Some people describe life as being anti-entropic, because living things take unorganized inorganic matter and order it in astonishingly intricate and complex ways. Is this not inconsistent with the law of increasing entropy? Not at all! No creature is an isolated system. In fact, every creature releases energy into its environment as it organizes matter within itself. This release causes the entropy of the creature's surroundings to increase *far* more than the creature decreases its own entropy by putting a few things in order.

Indeed, we should consider thoughts in a brain, movements of a muscle, growth, development, even evolution itself as examples of *entropy in action*. Sources of energy are tapped by living things and ultimately dissipated into the environment, and this natural, spontaneous flow of energy from being concentrated to dissipated is cleverly tapped by organisms to help accomplish just a little organization along the way. The beautiful order of life is *enabled by* and is thus a *manifestation* of the dissipation of energy (and consequent increase in entropy of the universe).

In some cases, the link with disorder is misleading

Figure T3.2
The contents of which glass have the higher entropy? (Credit: © McGraw-Hill Education/Chris Hammond, photographer)

Figure T3.3

Entropy in action (including the fish!). This is the wreck of the steamship Alicia in Key Biscayne National Park. (Credit: National Park Service)

Stars and galaxies form, the sun shines, tectonic plates move, storms rage, fish swim, and flowers bloom in the spring because all systems in the universe evolve toward macropartitions that maximize their entropy (see figure T3.3). Every physical process in our daily life, whether associated with growth or with decay, increasing order or increasing disorder, is not merely *consistent* with the second law of thermodynamics: it *expresses* that law.

The basic meaning of entropy is multiplicity, *not* disorder

There are some cases, therefore, for which the idea of *entropy as disorder* can be misleading. Whenever there appears to you to be a contradiction between the concepts of entropy as disorder and entropy as multiplicity, remember that *multiplicity* is the more basic concept.

T3.4   The Definition of Temperature

We saw in chapter T1 that Kelvin discovered how to define absolute temperature in a very fundamental and general way *without* referring to any particular thermoscope. My goal in this section is to show how Kelvin's definition of temperature follows from statistical mechanics.

Consider again our famous *paradigmatic thermal process*. A hot object is brought into contact with a cold object. Subsequently, heat flows from the hot object to the cold object, decreasing the energy (and thus entropy) of the hot object and increasing the energy (and thus entropy) of the cold object. Assuming that the two objects have a large number of molecules, we have seen the combined system evolve inexorably toward the most probable macropartition and subsequently remain there, in *thermal equilibrium*. The equilibrium macropartition will therefore be that macropartition with the greatest number of microstates, and so the greatest total entropy $S_{TOT} = S_A + S_B$.

Now, the combined system's entropy S_{TOT} is a function of the system's macropartition, which in a pure heat-transfer process (where the volume V, the number of molecules N, and other macroscopic characteristics of the interacting objects are held constant) is completely determined by the objects' energies U_A and U_B. Actually, we only need to know U_A to determine the macropartition, since $U_B = U - U_A$, where U is the fixed total energy of the combined system. The macropartition where S_{TOT} is maximum is specified by the value of U_A such that

$$0 = \frac{dS_{\text{TOT}}}{dU_A}. \tag{T3.7a}$$

(This is the usual way of finding the maximum of a function.) Since the system's total entropy $S_{\text{TOT}} = S_A + S_B$, we can rewrite this as follows:

$$0 = \frac{d(S_A + S_B)}{dU_A} = \frac{dS_A}{dU_A} + \frac{dS_B}{dU_A} \tag{T3.7b}$$

by the sum rule of differential calculus. Now, according to the chain rule,

$$\frac{dS_B}{dU_A} = \frac{dS_B}{dU_B}\frac{dU_B}{dU_A} = \frac{dS_B}{dU_B}(-1) \tag{T3.8}$$

since $U_B = U - U_A$. (More informally, note that because energy gained by A is lost by B, $dU_A = -dU_B$. Substituting this into dS_B/dU_A yields $-dS_B/dU_B$.) If we substitute equation T3.8 into equation T3.7b, we find that the two objects will be in thermal equilibrium if and only if

$$\frac{dS_A}{dU_A} = \frac{dS_B}{dU_B} \tag{T3.9}$$

The necessary condition for equilibrium

This tells us that the system's total entropy will be maximum when the slopes of the entropy curves of each subsystem are equal: see figure T3.4. When this is so, then the entropy gained by subsystem A when it gets a bit of energy dU_A from subsystem B is exactly balanced by the entropy that subsystem B loses in the process. If the slopes are *not* equal, the combined system can still increase its entropy by transferring energy from one subsystem to the other, so has not yet reached equilibrium.

Exercise T3X.1

Verify equation T3.9. (This is not hard.)

Note that we can calculate the quantities on either side of equation T3.9 *without reference to the other object*: we simply take the derivative of an object's entropy S with respect to its *own* thermal energy U (holding its other macroscopic properties constant). This equation thus tells us that when two objects are in the equilibrium macropartition, the quantity dS/dU calculated for each object must be the same.

But the fundamental meaning of temperature (as the zeroth law of thermodynamics asserts) is that the *temperature* of two objects is the same in equilibrium. Therefore, the quantities dS/dU and temperature T must be linked in some one-to-one relationship:

$$f(T) \equiv \frac{dS}{dU} \tag{T3.10}$$

where $f(T)$ is some as yet unknown function of the object's temperature T.

Different choices for $f(T)$ simply amount to different ways of linking dS/dU to numerical temperature values. *In principle, we could choose $f(T)$ to be anything we like as long as $f(T)$ has a unique value for every value of T* (so that two objects having the same temperature necessarily have the same value of dS/dU and vice versa): our choice would simply define a new temperature scale. So what function did Kelvin choose?

One feature we want in a temperature scale is that we want the *hotter* object in a pair (the one that gives up energy to the other) to have the *higher numerical temperature T*. As we have seen, energy flows spontaneously

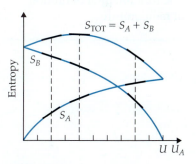

Figure T3.4
The system's total entropy is maximum where the slopes of the entropy curves for subsystems A and B are equal in magnitude and opposite in sign (at about $U_A = 0.4U$ in the case shown). The dashed lines for $U_A = 0.2U$ and $U_A = 0.8U$ show macropartitions where the slopes are not equal and the entropy is not maximum.

between two objects A and B in thermal contact because this allows the combined system's *total* entropy S_{TOT} to increase. Suppose that object A is the colder object, so that in an infinitesimal heat transfer process it gets a *positive* increment in energy dU_A from the other object. Since S_{TOT} must increase in this process, dS_{TOT}/dU_A must also be positive. Therefore, we have

$$0 < \frac{dS_{\text{TOT}}}{dU_A} = \frac{dS_A}{dU_A} + \frac{dS_B}{dU_A} = \frac{dS_A}{dU_A} - \frac{dS_B}{dU_B} \equiv f(T_A) - f(T_B) \tag{T3.11}$$

Exercise T3X.2

Explain the reasoning behind each step in in equation T3.11.

So for heat to flow from B to A, we must have $f(T_A) > f(T_B)$. But since object A here is the *cold* object, we see that as T increases, we want $f(T)$ to *decrease*. The *simplest* definition of $f(T)$ consistent with this requirement is $f(T) = 1/T$. So (following Kelvin) we define temperature T so that:

The fundamental definition of temperature

$$\frac{1}{T} \equiv \frac{dS}{dU} \text{ (with other variables constant)} \equiv \frac{\partial S}{\partial U} \tag{T3.12}$$

- **Purpose:** This equation defines an object's absolute temperature T in terms of the rate at which the object's entropy S increases with its internal energy U.
- **Limitations:** We must hold the object's other macroscopic variables (the number of molecules N, volume V, and so on) constant when evaluating the derivative (this is what the partial derivative notation $\partial S/\partial U$ means).

(You may have encountered **partial derivatives** already. If you have not, please don't be stressed by this new notation. A partial derivative is the same thing as an ordinary derivative, except that the curly ∂ symbols remind us to hold anything that S might depend on *except* for U constant when we evaluate the derivative with respect to U.)

Note that because we included Boltzmann's constant in the definition of entropy $S \equiv k_B \ln \Omega$, entropy has the same units as k_B which are J/K. Therefore, dS/dU has units of 1/K, ensuring that temperature T has units of K (kelvins). This is the main reason for including k_B in the definition of S.

Since in principle we can count the number of microstates available to an object in a given macropartition, and from that determine how $S \equiv k_B \ln \Omega$ varies with internal energy, we can calculate any object's temperature T from fundamental physical quantities. *This is a very important equation!* The rest of this unit will be almost exclusively focused on exploring its implications.

T3.5   Consistency with Historical Definitions

Equation T3.12 defines temperature T in a manner consistent with the convention that heat flows from objects with large values of T to objects with low values of T. However, this in itself does not ensure that the temperature scale so defined is consistent with historical scales such as the constant-volume gas thermoscope scale. In this section, we will see some evidence suggesting that these two scales are in fact equivalent.

We saw in chapter T1 that at sufficiently high temperatures, the thermal energy U of a monatomic solid satisfies the approximate relationship

$$\frac{dU}{dT} = 3Nk_B \tag{T3.13}$$

where T in this equation was consistent with temperature scales used historically. But we know how to calculate the multiplicity of an Einstein solid, so in principle we can use equation T3.12 to *calculate* $1/T$ for a monatomic solid (modeled as an Einstein solid) as predicted by the new definition. If we get a result consistent with the above, it will be strong evidence that our new definition of temperature coincides with older definitions.

If we start with equation T3.12, apply the definition $S = k_B \ln \Omega$, the chain rule, and the fact that the derivative of $\ln x$ is $1/x$, we get

$$\frac{1}{T} = \frac{\partial S}{\partial U} = k_B \frac{\partial}{\partial U}(\ln \Omega) = k_B \frac{1}{\Omega}\frac{\partial \Omega}{\partial U} = k_B \frac{1}{\Omega}\frac{\partial \Omega}{\varepsilon \partial q} \tag{T3.14}$$

where $q \equiv U/\varepsilon$ is the total number of fundamental energy units that the solid contains. According to equation T2.7, an Einstein solid's multiplicity is

$$\Omega = \frac{(q + 3N - 1)!}{q!(3N - 1)!} \tag{T3.15}$$

We can estimate the value of $\partial\Omega/\partial q$ by computing $\Delta\Omega/\Delta q$ for the smallest possible value of Δq (which is $\Delta q = 1$) while holding N fixed:

$$\frac{\partial\Omega}{\partial q} \approx \frac{\Delta\Omega}{\Delta q} = \frac{\Omega(q+1) - \Omega(q)}{1} = \frac{[(q+1) + 3N - 1]!}{(q+1)!(3N-1)!} - \frac{(q + 3N - 1)!}{q!(3N-1)!} \tag{T3.16}$$

But $(n + 1)! \equiv (n + 1) \cdot n \cdot (n - 1) \cdots 1 = (n + 1) \cdot n!$. If you substitute this into equation T3.16 and use equation T3.15, you should find that

$$\frac{\partial\Omega}{\partial q} \approx \frac{\Delta\Omega}{\Delta q} = \frac{q + 3N}{q + 1}\Omega - \Omega = \Omega\left[\frac{3N - 1}{q + 1}\right] \approx \Omega\left[\frac{3N}{q}\right] \tag{T3.17}$$

assuming both $3N$ and q are large compared to 1.*

Exercise T3X.3

Fill in the missing steps in equation T3.17.

If we substitute this back into equation T3.14 and use $q \equiv U/\varepsilon$, we find that according to our definition of temperature,

$$\frac{1}{T} = \frac{k_B}{\varepsilon}\frac{1}{\Omega}\left(\Omega\frac{3N}{q}\right) = k_B\frac{3N}{\varepsilon q} = \frac{3Nk_B}{U} \quad\Rightarrow\quad U = 3Nk_B T \tag{T3.18}$$

Taking the derivative of both sides of this with respect to T indeed yields the empirical result $dU/dT = 3Nk_B$. This strongly suggests that our definition of temperature coincides with older definitions. (We will see even more compelling evidence in chapter T5.)

---

*Technically, $\Delta\Omega/\Delta q$ will be a good approximation to $d\Omega/dq$ only if its value does not change significantly as q changes by $\Delta q = 1$. Note that according to equation T3.17, Ω increases by a factor of $(q + 3N)/(q + 1)$ as q increases by 1, and as $\Delta\Omega/\Delta q$ is proportional to Ω, it will increase by this factor also. So for the change in $\Delta\Omega/\Delta q$ to be small when $\Delta q = 1$, we must also have $q \gg 3N$, so that $(q + 3N)/(q + 1) \approx 1$. This means that this calculation works only when the solid is hot enough that each oscillator, on average, has many units of energy.

Example T3.3

Problem: When $q \gg 3N \gg 1$, one can show by an entirely different argument (see problem T3D.5) that an Einstein solid's multiplicity is roughly

$$\Omega(U, N) \approx \left(\frac{eU}{3N\varepsilon}\right)^{3N} \tag{T3.19}$$

where $e = 2.718$. Show that this equation also implies that $U \approx 3Nk_BT$.

Solution Equation T3.3 implies that

$$\ln \Omega = \ln\left(\frac{eU}{N\varepsilon}\right)^{3N} = 3N \ln \frac{eU}{N\varepsilon} = 3N \left(\ln e + \ln U - \ln N - \ln \varepsilon\right) \tag{T3.20}$$

The definition of temperature then implies that

$$\frac{1}{T} = \frac{\partial S}{\partial U} = \frac{\partial}{\partial U}\left(k_B \ln \Omega\right) \approx 3Nk_B \frac{\partial}{\partial U}\left(\ln e + \ln U - \ln N - \ln \varepsilon\right)$$

$$= 3Nk_B\left(0 + \frac{1}{U} + 0 + 0\right) = \frac{3Nk_B}{U} \tag{T3.21}$$

Multiplying both sides of this equation by UT yields $U \approx 3Nk_BT$.

T3.6   A Financial Analogy

My friend Daniel Schroeder, in his textbook *Thermal Physics* (Addison Wesley Longman, 2000), uses the following light-hearted analogy to make the definition of temperature clearer. An isolated system of objects that constantly exchange energy so as to maximize the system's total entropy is analogous to a community of people who exchange money so as to maximize not their own happiness but the community's total happiness. A *cold* object has a large value of dS/dU according to our definition of temperature, implying that its entropy increases rapidly as it receives energy and drops dramatically when it gives up energy. We might describe the analogous person whose happiness increases rapidly when receiving money and decreases rapidly when giving money as being "needy." Conversely, a hot object has a small value of dS/dU, implying that its entropy does not change much when it gains or loses energy. We would call the analogous person who is neither greatly excited by receiving money nor greatly perturbed to give it away "generous." If you put a generous person in contact with a needy person, the generous one will pay some money to the needy one to increase the community's happiness, just as a hot object will give up energy to a cold object to increase the system's entropy. Note also that judiciously redistributing money can increase a community's happiness (even though the total amount of money does not increase) in the same way that redistributing conserved energy among a system's parts can still increase its entropy. So we see that an analogy in which

The core of the analogy

Energy	$\leftrightarrow$	Money
Entropy	$\leftrightarrow$	Happiness
Temperature	$\leftrightarrow$	Generosity

can help us more intuitively understand the link between temperature, entropy, and energy (as long as we don't take it *too* seriously).

Now, a *normal* person (at least theoretically) becomes more generous as he or she receives money. Analogously, the temperature of a *normal* object (for example, an Einstein solid) becomes larger as it receives energy. If we draw a graph of the entropy S of a normal object versus its energy U, we get something like that shown in figure T3.5a: note that the slope $\partial S/\partial U$ decreases (implying that the temperature T increases) as U increases.

Types of people compared to types of systems

However, nothing in physics prevents an object's temperature from decreasing when you add energy. Indeed, systems (such as stars or star clusters) that are bound by gravity behave in this way. Adding energy to a star causes it to expand, and this ends up putting more energy into the gravitational potential energy of expansion than was added originally, requiring the star's temperature to decrease. We would call the analogous person who gets more needy as you give him or her money "miserly." Figure T3.5b shows a graph of S versus U for a miserly physical system. Fortunately, genuinely miserly people are rare in society, and miserly physical systems are also pretty rare.

In both normal and miserly systems, though, the system's entropy S *increases* when you add energy U. This is almost universally true, because more energy almost always means more ways to distribute that energy, meaning more microstates and thus greater entropy. For such systems $\partial S/\partial U$ and thus T are positive. Because $\partial S/\partial U$ cannot be larger than infinite, $T = 0$ (as Kelvin recognized) represents the absolute minimum value that T can have.

However, just as one very rarely encounters enlightened individuals whose happiness actually *increases* when they give money away, one very rarely encounters a physical system where S decreases as U increases (see figure T3.5c). Such an "enlightened" physical system has *negative* temperature according to our definition! Yet just as an enlightened person would gladly give away all his or her money and a normal person would be happy to receive it, such an "enlightened" physical system will give up all its energy to a normal system, and so is in that sense "hotter" than any normal object with positive temperature (which means that $T = 0$ is still the *coldest* that an object can be). Because we are so used to normal objects, this kind of a situation is extremely counterintuitive, but physical systems like this *can* be constructed with some effort. (Problem T3R.1 discusses an example of such a system.)

We will deal essentially entirely with "normal" physical objects in this course. Even so, it is good for your flexibility of mind to know that more exotic systems do exist. The financial analogy, even though it is a bit silly, helps one think clearly about *all* kinds of systems. We will also return to the financial analogy in chapter T8.

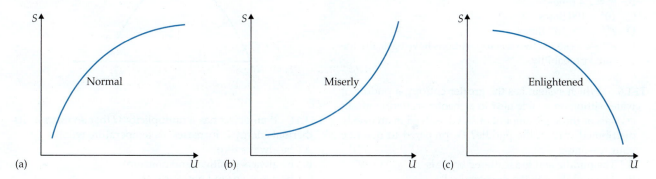

Figure T3.5
Entropy versus energy graphs for a normal object, a miserly object, and an enlightened object.

TWO-MINUTE PROBLEMS

T3T.1 A hot object is placed in contact with a cold object. We observe that heat flows spontaneously from the hot object to the cold object, but not in the other direction. According to the argument in this chapter, this is so because
A. this increases the entropies of both objects.
B. this decreases the entropy of the combined system.
C. the system will tend to evolve toward macropartitions that have more microstates.
D. *A* and *C*
E. *B* and *C*

T3T.2 Consider an object (object *A*) whose multiplicity is always 1, no matter how much energy you put into it. If you put a very large amount of energy into such an object, and place it into thermal contact with an Einstein solid (object *B*) having the same number of atoms but much less energy, what will happen? (*Hint:* Imagine a macropartition table for such a pair of objects.)
A. Energy flows from *A* to *B* until *A* has very little energy.
B. Energy flows from *B* to *A* until *B* has very little energy.
C. Energy flows from *A* to *B* until both objects have the same energy.
D. No energy will flow between *A* and *B* at all.
E. Something else happens (describe).

T3T.3 In a given macropartition, the total entropy of a system comprised of two or more parts is always equal to the *sum* of the entropies of those parts. T or F?

T3T.4 $10^{62}/10^{60} = 100$. What is $\ln 10^{62}/\ln 10^{60}$? (Select the closest response.)
A. 100
B. 10
C. 4.6
D. 1.0

T3T.5 The entropy of a certain macropartition of a combined system is $102k_B$. The entropy of another macropartition is $204k_B$. How much more likely is the system to be in the second macropartition than the first? (Select the closest response.)
A. 2 times
B. $e^2 = 7.4$ times
C. $10^2 = 100$ times
D. $e^{102} = 2 \times 10^{44}$ times
E. $e^{102k_B} \approx e^0$; so the two macropartitions have basically the same probability

T3T.6 Which system has the greater entropy, a puddle of water sitting on a table next to a chaotic scattering of salt crystals or the same amount of salt dissolved in an evenly distributed way in the puddle? Be prepared to describe your reasoning.
A. The puddle and the scattered crystals.
B. The puddle with the dissolved salt.
C. Both have the same entropy.
D. One needs detailed models of water and salt to say.

T3T.7 Object *A*'s entropy increases quite a bit when we give it a certain tiny amount of energy, whereas object *B*'s entropy increases by a much smaller amount with the same gift. Which has the higher temperature?
A. Object *A*
B. Object *B*
C. One cannot say without more information.

T3T.8 As a normal object's thermal energy *U* goes to zero, the slope of a graph of its entropy *S* versus *U*
A. approaches infinity.
B. approaches some positive value less than infinity.
C. approaches zero.
D. approaches some finite negative value.
E. approaches negative infinity.
F. depends entirely on the object's properties.

T3T.9 Consider a system whose multiplicity is 1, no matter how much energy one puts into it. What is such a system's temperature?
A. Zero
B. Positive
C. Negative
D. Undefined

T3T.10 Suppose that an object's entropy increases from zero when its thermal energy is $U = 0$ to some maximum value when $U = \frac{1}{2}U_{max}$, and then goes back to zero when $U = U_{max}$ (as shown in the drawing below). The greatest thermal energy *U* that such an object could have when in thermal equilibrium with a *normal* object is closest to which of the following?
A. 0
B. $\frac{1}{4}U_{max}$
C. $\frac{1}{2}U_{max}$
D. $\frac{3}{4}U_{max}$
E. U_{max}
F. Other (specify)

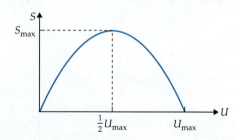

T3T.11 If an object has a multiplicity Ω that *decreases* as its thermal energy *U* increases, its temperature would
A. be always zero.
B. be always positive (and nonzero).
C. be negative (and nonzero).
D. increase as *U* increases.
E. decrease as *U* increases.

HOMEWORK PROBLEMS

Basic Skills

T3B.1 An Einstein solid in a certain macrostate has a multiplicity of 3.8×10^{280}. What is its entropy (expressed as a multiple of k_B)?

T3B.2 Two interacting Einstein solids in a certain macropartition have multiplicities of 4.2×10^{320} and 8.6×10^{132}.
(a) What are the entropies of each solid?
(b) What is the combined system's entropy in this macropartition?
(Express all entropies as multiples of k_B.)

T3B.3 Two interacting Einstein solids in a certain macropartition have multiplicities of 8.9×10^{1032} and 2.1×10^{1086}.
(a) What are the entropies of each solid?
(b) What is the combined system's entropy in this macropartition?
(Express entropies as multiples of k_B.)

T3B.4 (Calculating Einstein solid entropies.)
(a) What is the entropy of an Einstein solid with 5 atoms and an energy of 15ε?
(b) What is the entropy of an Einstein solid with 50 atoms and an energy of 100ε? (*Hint:* I recommend using Stat-Mech to calculate the multiplicity.)
(Express your answers as multiples of k_B.)

T3B.5 Objects A and B have different temperatures and initial entropies of 22 J/K and 47 J/K. We bring the objects into thermal contact and allow them to come to equilibrium. What is the most that we can say about the combined system's entropy at this point?

T3B.6 Is it *really* true that the entropy of an isolated system consisting of two Einstein solids *never* decreases? (Consider a pair of very small solids.) Why is this statement more accurate for large systems than for small systems? Explain in your own words.

T3B.7 We measure an object's entropy to increase by 0.1 J/K when we add 35 J of energy. What is its approximate temperature? (Assume that the object's temperature does not change much when we add the energy.)

T3B.8 Suppose that we increase an object's energy by 10 J while its temperature remains roughly constant at 20°C. By how much does the object's entropy increase?

T3B.9 A certain Einstein solid's entropy changes from $305.2k_B$ to $338.1k_B$ when we add one unit ε of energy.
(a) What is the value (and units) of $k_B T / \varepsilon$ for this solid?
(b) If $\varepsilon = 1.0$ eV, what is its temperature T?

T3B.10 Does it make sense to talk about the temperature of a vacuum? If so, how could you measure or calculate it? If not, why not?

Modeling

T3M.1 You ask your roommate to clean up a mess that your roommate made in your room. Your roommate refuses, because cleaning up the mess would violate the second law of thermodynamics, and campus security's record of your roommate's legal violations is already excessive. Gently but firmly explain why complying will not put your roommate at risk of such an infraction.

T3M.2 The classic statement of Murphy's law reads, "If something can go wrong, it will." Explain how this is really a consequence of the second law of thermodynamics. (*Hint:* What is the entropy of "wrong" in a given context compared to the entropy of "right"?)

T3M.3 A certain macropartition of two Einstein solids has an entropy of $305.2k_B$. The next macropartition closer to the most probable one has an entropy of $335.5k_B$. If the system is initially in the first macropartition and we check it again later, how many times more likely is it to have moved to the other than to have stayed in the first?

T3M.4 The entropy of the most probable macropartition for a certain system of Einstein solids is $6025.3k_B$, while the entropy of an extreme macropartition is only $5755k_B$. What is the probability of finding the system at a given time in the extreme macropartition compared to that of finding it in the most probable macropartition? (*Hint:* See problem T3D.3 for some possibly useful tricks.)

T3M.5 Suppose that a certain substance's entropy as a function of N and U is given by the formula $S = Nk_B \ln U$. Using the definition of temperature, show that the thermal energy of this substance is related to its temperature by the expression $U = Nk_B T$.

T3M.6 Suppose that a certain solid's multiplicity is given by $\Omega(U, N) = Ne^{\sqrt{NU/\varepsilon}}$, where ε is some unit of energy.
(a) How would the energy of an object made out of this substance depend on its temperature?
(b) Would this be a "normal" substance in the sense defined in section T3.6?

T3M.7 The multiplicity of an ideal monatomic gas with N atoms, thermal energy U, and volume V turns out to be roughly $\Omega(U, V, N) = CV^N U^{3N/2}$, where C is a constant that depends on N alone. Use this information and the definition of temperature to determine how the thermal energy of a monatomic gas depends on its temperature.

T3M.8 Consider an Einstein solid having $N = 20$ atoms.
(a) What is the solid's temperature when it has an energy of 10ε, assuming that $\varepsilon = \hbar\omega \approx 0.02$ eV (a physically reasonable value for a solid)? Calculate this directly from the definition of temperature by finding S at 10ε and 11ε, computing $dS/dU \approx [S(11\varepsilon) - S(10\varepsilon)]/\varepsilon$, and then applying the definition of temperature. (You may find that your work will go faster if you use StatMech to tabulate the multiplicities.)
(b) How does this compare with the result from the formula $U = 3NK_BT$ (which is only accurate if N is large and $U/3N\varepsilon \gg 1$)?
(c) Repeat for $N = 200$ and $U = 100\varepsilon$. (*Hint:* See problem T3D.3 for some potentially useful tricks.)

T3M.9 One of the important things about the definition of temperature in equation T3.12 is that we can use it to *infer* the entropies (and thus the multiplicities) for objects even when we can't easily construct a detailed model that would allow us to calculate those multiplicities directly. Consider the following examples.
(a) Melting a gram of ice at 0°C requires adding 333 J of heat energy. How much *must* the ice's entropy increase as it melts?
(b) A typical person dissipates about 60 W of energy into a room at 20°C even when just sitting. Assuming that the room's temperature doesn't increase very much while absorbing this energy, by about how much *must* absorbing this person's energy increase the room's entropy in an hour?
We will do more calculations of this type in chapter T8.

Derivation

T3D.1 *In principle*, the entropy of an isolated system decreases a little bit whenever random processes cause its macropartition to fluctuate away from the most probable macropartition. Large fluctuations can certainly happen in small systems (see figure T2.5a). But is this really a possibility for a typical macroscopic system? Suppose that we can measure the entropy of a system of two solids to within 2 parts in 1 billion. This means that we could just barely distinguish a system that has an entropy of 4.99999999 J/K (eight 9s!) from one that has 5.00000000 J/K. (This is a reasonable entropy for a macroscopic system.)
(a) Suppose that the entropy of the equilibrium macropartition is 5.00000000 J/K. Show that the approximate probability of finding the system in a macropartition with entropy 4.99999999 J/K (that is, with an entropy that is only barely measurably smaller) at any given later time is about $10^{315,000,000,000,000}$ times smaller than the probability we will still find it to have an entropy of 5.00000000 J/K. (*Hint:* See problem T3D.3.)

(b) Defend the statement that the entropy of an isolated system in thermal equilibrium *never* decreases.

T3D.2 We define the natural logarithm function $\ln x$ (where x is positive) to be the inverse of the exponential function: $e^{\ln x} \equiv x$. Use this definition to prove the following:
(a) Prove that $\ln(xy) = \ln x + \ln y$.
(b) Prove that $\ln(x^y) = y \ln x$.
(c) Prove that

$$\frac{d}{dx}\ln x = \frac{1}{x} \qquad (T3.22)$$

(*Hint:* Use the chain rule to take the derivative of $e^{\ln x}$.)
(d) Prove the very useful approximation

$$\ln(1 + x) \approx x \quad \text{when } x \ll 1 \qquad (T3.23)$$

[*Hint:* We can write the exponential as a power series: $e^x = 1 + x + x^2/2! + x^3/3! + \cdots$. Alternatively, you might write a Taylor series expansion of $\ln(1 + x)$.]

T3D.3 Tricks for dealing with ginormous numbers.
(a) Use the results of problem T3D.2 to prove that

$$\ln(a \times 10^b) = \ln a + b \ln 10 \qquad (T3.24)$$

(b) Use $e^{\ln x} \equiv x$ (the definition of the logarithm function) and the fact that $e^{ab} = (e^a)^b$ to prove that

$$e^x = 10^{x/\ln 10}. \qquad (T3.25)$$

T3D.4 In section T3.4, I argued on fairly fundamental grounds that $dS/dU = f(T)$. In principle, we could define $f(T)$ to be anything that we like: this would amount to *defining* temperature and its scale. Still, *some* definitions would violate deeply embedded preconceptions about the nature of temperature. For example, the *simplest* definition of temperature would be $dS/dU = T^*$.
(a) Show that this definition would imply that T^* has units of inverse kelvins.
(b) Prove from first principles that this definition would imply that heat would flow spontaneously from objects with low T^* to objects with high T^*. This would imply that objects with low values of T^* are *hot*, while objects with high values of T^* are *cold* (we might want to call T^* so defined as *coolness* instead of *temperature*). While we *could* define temperature in this way, it would really fly in the face of convention (if not intuition).
(c) If we define coolness T^* this way, what is the ordinary temperature T of a totally uncool object (at $T^* = 0$)? What about an object that is infinitely cool ($T^* = \infty$)?
(d) One of the counterintuitive things about the ordinary temperature scale is that negative temperatures are hotter than positive temperatures. Is this a problem with our coolness scale as well?

T3D.5 When N is large, an Einstein solid's multiplicity is

$$\Omega(U,N) = \frac{(q + 3N - 1)!}{q!(3N - 1)!} \approx \frac{(q + 3N)!}{q!(3N)!} \qquad (T3.26)$$

where $q = U/\varepsilon$. Now, **Stirling's approximation** asserts that

$$\ln(m!) \approx m \ln m - m \quad \text{when } m \gg 1 \qquad \text{(T3.27)}$$

We also know (see problem T3D.2d) that

$$\ln(1 + x) \approx x \quad \text{when } x \ll 1 \qquad \text{(T3.28)}$$

(a) In the limit that $U/(3N\varepsilon) \gg 1$ (that is, the solid has enough energy to give each atom many units ε of energy), use the approximations above to show that

$$\ln \Omega \approx 3N \ln\left(\frac{U}{3N\varepsilon}\right) + 3N \qquad \text{(T3.29)}$$

(b) Show that taking the exponential of both sides yields

$$\Omega \approx \left(\frac{eU}{3N\varepsilon}\right)^{3N} \qquad \text{(T3.30)}$$

as claimed in equation T3.19.

Rich-Context

T3R.1 Suppose that a certain kind of atom can be in one of only two quantum states, one with zero energy and the other with energy ε. Consider an object with N such atoms. Note that such an object will have a thermal energy U such that $0 \leq U \leq N\varepsilon$.

(a) Argue that the object's multiplicity is

$$\Omega(U, N) = \frac{N!}{q!(N - q)!} \quad \text{where } q = \frac{U}{\varepsilon} \qquad \text{(T3.31)}$$

(Hint: Argue that this situation is analogous to drawing q red and $N - q$ green marbles from a bag. How many unique patterns can we create?)

(b) Adapt the argument in section T3.5 to show that

$$\frac{d}{dq} \ln \Omega \approx \frac{N - 2q}{q} \qquad \text{(T3.32)}$$

when both $N \gg 1$ and $q \gg 1$.   (continues)

(c) Find this object's temperature as a function of U.
(d) Is this object a "normal" object if $U < \frac{1}{2}N\varepsilon$? Explain.
(e) Is this object a "normal" object if $U > \frac{1}{2}N\varepsilon$? Explain.
Comment: This model is actually a decent description of certain kinds of solids when we place them in a strong magnetic field at very low temperatures. Under such circumstances, a given atom's magnetic moment can be either aligned with the field (which we can take to be the zero-energy state) or anti-aligned with the field (which we can take to be the excited state).

T3R.2 According to a checkout-counter news source, space aliens give top scientists an object made of a substance that can store thermal energy but whose multiplicity actually *decreases* as its energy increases. Answer the following questions *qualitatively*, but carefully, supporting your answers with arguments based on the ideas in this chapter. (Hint: What might a macropartition table look like for such an object in contact with an Einstein solid?)

(a) What will happen to the energy in such an object if we place it in thermal contact with a normal object (such as an Einstein solid)?
(b) Can such an object ever be in thermal equilibrium with a normal object?
(c) Will putting such an object in a flame warm it?
(d) How might you increase the thermal energy of such an object?
(e) Do you think that we can assign a meaningful temperature to such an object? Why or why not?
Comment: This is not just a science fiction scenario: certain real physical systems can exhibit such behavior. For example, if one sets up the system described in problem T3R.1 with enough initial energy in a very low-temperature environment, it can behave in this way. In general, this kind of weird behavior is possible in systems where there is an upper limit on the energy that a molecule can hold, which is not the case in Einstein solids or most other substances.

ANSWERS TO EXERCISES

T3X.1 Applying equation T3.8 to equation T3.7b yields

$$0 = \frac{dS_A}{dU_A} + \frac{dS_B}{dU_A} = \frac{dS_A}{dU_A} - \frac{dS_B}{dU_B} \qquad \text{(T3.33)}$$

Adding dS_B/dU_B to both sides yields

$$\frac{dS_A}{dU_A} = \frac{dS_B}{dU_B} \qquad \text{(T3.34)}$$

T3X.2 Equation T3.11 (repeated here) says that

$$0 < \frac{dS_{\text{TOT}}}{dU_A} = \frac{dS_A}{dU_A} + \frac{dS_B}{dU_A} = \frac{dS_A}{dU_A} - \frac{dS_B}{dU_B} \equiv f(T_A) - f(T_B)$$

The first equality follows from equation T3.7b, the second from substituting the result of equation T3.8, and the third from the definition of $f(T)$ given in equation T3.10.

T3X.3 Because $(n + 1)! = (n + 1) \cdot n!$, it follows that

$$\frac{[(q + 1) + 3N - 1]!}{(q + 1)!(3N - 1)!} = \frac{(q + 1 + 3N - 1)(q + 3N - 1)!}{(q + 1)q!(3N - 1)!}$$

$$= \frac{q + 3N}{q + 1} \frac{(q + 3N - 1)!}{q!(3N - 1)!} = \frac{q + 3N}{q + 1} \Omega \qquad \text{(T3.35)}$$

This explains the first equality. Extracting the common factor of Ω in equation T3.16 and putting things over a common denominator, yields

$$\frac{q + 3N}{q + 1} \Omega - \Omega = \Omega\left[\frac{q + 3N}{q + 1} - 1\right]$$

$$\approx \Omega\left[\frac{q + 3N - (q + 1)}{q + 1}\right] = \Omega\left[\frac{3N - 1}{q + 1}\right] \qquad \text{(T3.36)}$$

In the last step in equation T3.17, we are assuming that $3N$ and q are both so large that we can neglect the ones in the numerator and denominator.

T4

The Boltzmann Factor

Chapter Overview

Introduction
This chapter introduces a very powerful tool for analyzing thermal systems that we will find very useful in both this chapter and the next.

Section T4.1: The Boltzmann Factor
We define a (thermal) **reservoir** to be an object so large that it can provide or absorb any energy likely to be exchanged in a situation of interest without undergoing a significant change in its temperature. Consider a small quantum system in thermal contact with such a reservoir. The probability that the quantum system will be in a quantum state with energy E is proportional to the multiplicity of the combined system. But the multiplicity of the quantum system in a given quantum state is 1 by definition, so the multiplicity of the combined system is the same as the reservoir's multiplicity. The reservoir's multiplicity will decrease as the quantum system's energy E increases because that energy comes at the expense of the reservoir's energy. A short calculation involving the definition of temperature implies that the *reservoir's* multiplicity (and thus the state's probability) in fact decreases *exponentially* with E:

$$\Pr(E) = \frac{1}{Z} e^{-E/k_B T} = \frac{e^{-E/k_B T}}{\displaystyle\sum_{\text{all states}} e^{-E_i/k_B T}} \tag{T4.8}$$

- **Purpose:** This equation describes the probability that a small system in thermal contact with a reservoir at absolute temperature T will be in a quantum state (that is, a microstate) with energy E, where E_i is the energy of the ith small-system quantum state, Z is a constant of proportionality called the **partition function,** and k_B is Boltzmann's constant.
- **Limitations:** The reservoir must be large enough that it can provide the small system with any energy it is likely to have without suffering a significant change in its temperature T.
- **Notes:** We call $e^{-E/k_B T}$ the **Boltzmann factor.**

Section T4.2: Some Simple Applications
Equation T4.8 has a large number of useful applications. This section illustrates how we can use the equation to calculate the probabilities of the two different configurations of a certain molecule in a solution at room temperature, and the probability that a hydrogen atom's electron will be in its first excited state at room temperature.

Section T4.3: The Average Energy of a Quantum System
We can calculate the average energy of a general quantum system as follows:

$$E_{\text{avg}} = E_0 \Pr(E_0) + E_1 \Pr(E_1) + E_2 \Pr(E_2) + \cdots = \sum_{\text{all } n} E_n \Pr(E_n) \tag{T4.23}$$

If the quantum system is in contact with a reservoir at temperature T, the nth quantum state's probability is $Z^{-1}e^{-E_n/k_BT}$, so

$$E_{avg} = \sum E_n \left(\frac{e^{-E_n/k_BT}}{Z} \right) = \frac{\sum E_n e^{-E_n/k_BT}}{\sum e^{-E_n/k_BT}} \qquad (T4.25)$$

- **Purpose:** This equation describes how we can calculate the average energy E_{avg} of a quantum system in thermal contact with a reservoir at temperature T, where E_n is the energy of the quantum system's nth state, k_B is Boltzmann's constant, and $Z = \sum e^{-E_n/k_BT}$. The sums are over *all* the system's states.
- **Limitations:** This equation only works for a system in contact with something large enough to be considered a reservoir for the energies the system might typically have.

If we know a quantum system's partition function Z as a function of temperature T, then the expression above is equivalent to

$$E_{avg} = -\frac{1}{Z}\frac{dZ}{d\beta}, \quad \text{where } Z \equiv \sum e^{-E_n\beta} \text{ and } \beta \equiv \frac{1}{k_BT} \qquad (T4.34)$$

- **Purpose:** This equation describes a quick method for calculating a quantum system's average energy E_{avg} if you know its partition function Z as a function of temperature T (k_B is Boltzmann's constant). The sum is over all energy states of the quantum system.
- **Limitations:** This applies only to a quantum system in thermal contact with a reservoir at temperature T.

Section T4.4: Application to the Einstein Solid

If we apply these techniques to a single oscillator in an Einstein solid (treating the rest of the solid as a reservoir), we can calculate the average energy per oscillator, and therefore the total thermal energy U of an Einstein solid containing $N \gg 1$ atoms and the more easily measured **heat capacity** dU/dT of that solid. The results are

$$U = \frac{3N\varepsilon}{e^{\varepsilon/k_BT} - 1} \quad \text{and} \quad \frac{dU}{dT} = 3Nk_B \left[\frac{e^{1/2\tau}}{\tau(e^{1/\tau} - 1)} \right]^2 \text{ where } \tau = \frac{k_BT}{\varepsilon} \qquad (T4.41, 42)$$

In the last chapter, we could only calculate $dU/dT = 3Nk_B$ in the high-temperature limit (which we verify here), but here we have an expression for dU/dT valid for *all* temperatures. This expression approximately matches experimental results (though various discrepancies illustrate where the Einstein model is inadequate or incomplete). In particular, this result illustrates that the heat capacity goes exponentially to zero as $T \ll T_E \equiv \varepsilon/k_B$ (the **Einstein temperature for the solid**), because at such temperatures, exciting an oscillator even to the $n = 1$ level (which is necessary if the oscillator is to store any thermal energy at all) is exceedingly improbable.

Please note that the "Einstein temperature" for a given solid is simply the temperature T_E at which the thermal energy k_BT that is typically available to a quantum system happens be equal to the separation ε between the energy levels of that solid's oscillators: $k_BT_E = \varepsilon$. T_E therefore represents a critical temperature for that solid against which other temperatures can be compared. In particular, when $T \gg T_E$, the solid's heat capacity dU/dT approaches $3Nk_B$, but when $T \ll T_E$, dU/dT goes to zero (the solid's ability to store thermal energy is "frozen out").

Comparing equation T4.42 with the experimental data actually provides a way to estimate the effective spring constant of the forces holding an atom in a monatomic crystal: for iron, the effective spring constant is 170 N/m.

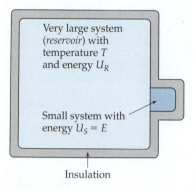

Figure T4.1

A very small system in thermal contact with a large system (both isolated from everything else).

Definition of a *reservoir*

Derivation of the Boltzmann factor

Here is where the reservoir assumption enters

T4.1   The Boltzmann Factor

Much of the rest of this text will be devoted to exploring consequences of the fundamental equation $1/T = \partial S/\partial U$. In this chapter, we will be developing some very useful tools that will help us with systems other than Einstein solids.

Consider a small system in thermal contact with a very much larger system (see figure T4.1 for a schematic diagram). For example, the small system could be a single gas molecule in contact with the rest of the gas in a room, or a single solute molecule in a solution. Our task is to answer the following question: When a small system is in contact with a large system with a given temperature T, how does the probability that the small system will be in a given quantum state (or microstate) depend on that state's energy E?

We can answer this question fairly easily in the limit that the large system is *very* much larger than the small system. We have seen before that the change dU in an object's energy is related to its change in temperature dT by the equation $dU = mc\,dT$, where m is the object's mass and c is its specific heat. So if a system is sufficiently massive, it can supply or absorb a significant amount of thermal energy while suffering only a negligible change in temperature. A **reservoir** is a system so large that it can absorb or supply any energy of interest in a given situation without undergoing a measurable change in temperature. In what follows, we will assume that the large system is large enough to be a reservoir with respect to the typical energies that the small system might absorb or emit.

Now consider two quantum states of the small system with energies E_0 and $E_1 > E_0$, respectively. What is the probability that the small system will be in the state with energy E_1 relative to that for the state with E_0? The fundamental assumption of statistical mechanics implies that the ratio of the probabilities is equal to the ratio of the combined system's multiplicities:

$$\frac{\Pr(E_1)}{\Pr(E_0)} = \frac{\Omega_{SR,1}}{\Omega_{SR,0}} \tag{T4.1}$$

where $\Omega_{SR,0}$ and $\Omega_{SR,1}$ are the combined system's multiplicities when the small system has energies E_0 and E_1, respectively. In general, the combined-system multiplicity is the product of the multiplicities of both subsystems: $\Omega_{SR} = \Omega_S\Omega_R$. However, in both cases, we are interested in the probability of the small system being in a *single* quantum state, so $\Omega_S = 1$ by definition. So

$$\frac{\Pr(E_1)}{\Pr(E_0)} = \frac{1 \cdot \Omega_{R1}}{1 \cdot \Omega_{R0}} = \frac{\Omega_{R1}}{\Omega_{R0}} \tag{T4.2}$$

where Ω_{R1} is the *reservoir's* multiplicity when the *small* system has energy E_1 and so on. Now, the reservoir's entropy, by definition, is $S_R = k_B \ln \Omega_R$, so $\Omega_R = e^{S_R/k_B}$. Plugging this into equation T4.2, we get

$$\frac{\Pr(E_1)}{\Pr(E_0)} = \frac{e^{S_{R1}/k_B}}{e^{S_{R0}/k_B}} = e^{(S_{R1}-S_{R0})/k_B} = e^{\Delta S_R/k_B} \tag{T4.3}$$

where ΔS_R is the change in the *reservoir's* entropy when the small system's energy increases from E_0 to E_1. If the combined system is isolated from the rest of the world, though, then the the combined system's total energy must be conserved. This means that if the small system's energy increases by $\Delta U_S = E_1 - E_0$, the energy of the reservoir *decreases* by the same amount:

$$\Delta U_R = -(E_1 - E_0) \tag{T4.4}$$

Finally, note that the fundamental equation $1/T = \partial S/\partial U$ tells us that if the small system's volume does not significantly change when its energy increases, and the reservoir is large enough that changing its energy by ΔU_R does not change its temperature T measurably, then $\Delta S_R \approx \Delta U_R / T$. Substituting these results into equation T4.3 yields

$$\frac{\text{Pr}(E_1)}{\text{Pr}(E_0)} = e^{\Delta U_R / k_B T} = e^{-(E_1 - E_0)/k_B T} = \frac{e^{-E_1/k_B T}}{e^{-E_0/k_B T}} \tag{T4.5}$$

This relationship must be true for *all* pairs of small-system quantum states. Consequently, the probability that a small system in contact with a reservoir at temperature T will be in a quantum state with any energy E must be

$$\text{Pr}(E) \propto e^{-E/k_B T} \quad \text{or} \quad \text{Pr}(E) = \frac{1}{Z} e^{-E/k_B T} \tag{T4.6}$$

where $1/Z$ is a constant of proportionality that must (by definition) be the same for all quantum states of the small system.

Since the *total* probability of the small system being in *some* quantum state must be 1, we must have

$$1 = \sum_{\text{all states}} \text{Pr}(E_i) = \sum_{\text{all states}} \frac{1}{Z} e^{-E_i/k_B T} = \frac{1}{Z} \sum_{\text{all states}} e^{-E_i/k_B T}$$

$$\Rightarrow \quad Z = \sum_{\text{all states}} e^{-E_i/k_B T} \tag{T4.7}$$

where E_i is the energy of the ith small-system quantum state. The absolute probability of the small-system being in a quantum state with energy E is thus

$$\text{Pr}(E) = \frac{1}{Z} e^{-E/k_B T} = \frac{e^{-E/k_B T}}{\sum\limits_{\text{all states}} e^{-E_i/k_B T}} \tag{T4.8}$$

- **Purpose:** This equation describes the probability that a small system in thermal contact with a reservoir at absolute temperature T will be in a quantum state (that is, a microstate) with energy E, where E_i is the energy of the ith small-system quantum state, Z is a constant of proportionality called the **partition function**, and k_B is Boltzmann's constant.
- **Limitations:** The reservoir must be large enough that it can provide the small system with any energy it is likely to have without suffering a significant change in its temperature T.
- **Notes:** We call $e^{-E/k_B T}$ the **Boltzmann factor.**

The probability that the small system is in quantum state with energy E

This important equation has *many* interesting applications. Note that it follows directly from the definitions of entropy and temperature, the fundamental assumption that all microstates of the combined system are equally probable, and the assumption that the reservoir's temperature is essentially unchanged as the small system's energy fluctuates.

Equation T4.8 implies that the probability of a small system being in a quantum state with energy E decreases exponentially as E increases. This makes sense if you recall that increasing the small system's energy means decreasing the reservoir's energy, which decreases the number of microstates available to the reservoir. As the probability of the small system's having energy E in this situation is entirely determined by the number of microstates available to the *reservoir* (see equation T4.2), the probability *should* decrease as the small system's energy increases. Equation T4.8 asserts that this probability, in fact, decreases exponentially (see figure T4.2).

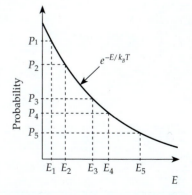

Figure T4.2
The probability of a small system's quantum state decreases exponentially as the state's energy E increases.

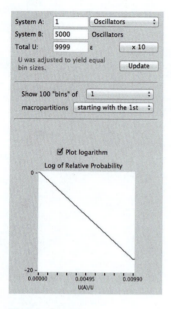

Figure T4.3
The fact that a graph of $\ln \Omega = S/k_B$ versus $E = U_A$ is a straight line indicates that Ω decreases exponentially with E. [The vertical scale on the graph actually indicates $\log_{10} \Omega = \ln \Omega / \ln(10)$.]

We can display this exponential dependence for systems of quantum oscillators using the StatMech application introduced in chapter T2. First, note that the probability that the combined system is in a macropartition where the small system is in a quantum state with energy E is Ω_{SR}/Ω_{TOT}, where Ω_{SR} is the macropartition's multiplicity and Ω_{TOT} is the total number of microstates available to the combined system. Therefore, equation T4.8 quite generally implies that

$$\frac{S}{k_B} \equiv \ln \Omega_{SR} = \ln [\Omega_{TOT} \Pr(E)] = \ln \left(\frac{\Omega_{TOT}}{Z} e^{-E/k_B T} \right) = \text{constant} - \frac{E}{k_B T} \quad \text{(T4.9)}$$

where $S \equiv k_B \ln \Omega_{SR}$ is the entropy of the combined system in that macropartition. So if we plot S/k_B versus the small system's energy E, we should get a straight line with slope $-1/k_B T$.

To get StatMech to display such a graph, we need to do several things. First, we need to set up solid A so that it consists of a single *oscillator* (not a three-oscillator "atom"). This ensures that solid A's multiplicity is always 1. We can do this by selecting "Oscillators" instead of "Atoms" in the menu to the right of the text fields for N_A and N_B and then typing 1 for N_A. Solid B will be the reservoir, so type in a fairly large number here, say 5000, so that it is *much* larger than the small system.

Second, we want to display a graph of S/k_B versus $E = U_A$, and we are most interested in the part of the graph where $U_A \ll U_B$, because if system A ever steals a significant fraction of system B's energy, B's temperature will change and therefore we cannot consider it to be a good reservoir. If we set the menus to display 100 "bins" of 1 macropartition starting with the first and set $U = 9999\varepsilon$, then we will be focusing on only those macropartitions where the small system's energy is less than 1% of U. If we click the "Plot logarithm" check, we will see a plot of $\log_{10} \Omega_{SB}$ for these macropartitions. Because $\log_{10} x = \ln x / \ln 10$, $\log_{10} \Omega_{SR} = S/(k_B \ln 10)$, so the function displayed is proportional to S as a function of $E = U_A$. Figure T4.3 shows the resulting graph. We see that the graph is indeed a straight line with negative slope.

Exercise T4X.1

Suppose that the difference between oscillator energy levels is $\varepsilon = 0.01$ eV. From the slope of the graph in figure T4.3, estimate the reservoir's temperature when it has an internal energy of nearly $10,000\varepsilon$.

T4.2   Some Simple Applications

Equation T4.8 has a huge range of applications that we can only begin to explore in this chapter. In this section, we will consider a few illustrative examples.

Example T4.1

Problem: Suppose that a certain molecule has two configurations, one with an energy E_0 and another with energy $E_1 = E_0 + 0.051$ eV. If we have N molecules in a solution at room temperature, roughly what fraction will be in each configuration?

Solution We will assume the solution here is large enough compared to a single molecule that it qualifies as a reservoir (this should be an excellent approximation) and that each configuration corresponds to a unique quantum state for the molecule. We will also assume that room temperature is $T = 22°C = 295$ K. Given these assumptions, we can in principle use equation T4.8 to compute the probability of each configuration. For example,

$$\text{Pr}(E_0) = \frac{1}{Z} e^{-E_0/k_B T} \tag{T4.10a}$$

where

$$Z \equiv \sum_n e^{-E_n/k_B T} = e^{-E_0/k_B T} + e^{-E_1/k_B T} \tag{T4.10b}$$

We could calculate this fairly easily if we knew E_0, but we do not. However, we do know that $E_1 = E_0 + 0.051$ eV. If we pull out a factor of $e^{-E_0/k_B T}$ from the sum for Z, we find that

$$Z = e^{-E_0/k_B T}\left(1 + \frac{e^{-E_1/k_B T}}{e^{-E_0/k_B T}}\right) = e^{-E_0/k_B T}(1 + e^{-(E_1-E_0)/k_B T}) \tag{T4.11}$$

Substituting this into equation T4.10, yields

$$\text{Pr}(E_0) = \frac{e^{-E_0/k_B T}}{e^{-E_0/k_B T}(1 + e^{-(E_1-E_0)/k_B T})} = \frac{1}{1 + e^{-(E_1-E_0)/k_B T}} \tag{T4.12}$$

In our particular case,

$$k_B T = (1.38 \times 10^{-23} \text{ J/K})(295 \text{ K})\left(\frac{1 \text{ eV}}{1.60 \times 10^{-19} \text{ J}}\right) = 0.0254 \text{ eV} \tag{T4.13}$$

which in turn implies that

$$\frac{E_1 - E_0}{k_B T} = \frac{0.051 \text{ eV}}{0.0254 \text{ eV}} = 2.0 \tag{T4.14a}$$

$$\text{Pr}(E_0) = \frac{1}{1 + e^{-(E_1-E_0)/k_B T}} = \frac{1}{1 + e^{-2.0}} = 0.88 \tag{T4.14b}$$

We can calculate the probability of the other configuration in the same way:

$$\text{Pr}(E_1) = \frac{e^{-E_1/k_B T}}{Z} = \frac{e^{-E_1/k_B T}}{e^{-E_0/k_B T}(1 + e^{-(E_1-E_0)/k_B T})}$$

$$= \frac{e^{-(E_1-E_0)/k_B T}}{1 + e^{-(E_1-E_0)/k_B T}} = \frac{e^{-2.0}}{1 + e^{-2.0}} = 0.12 \tag{T4.15}$$

Note that these two probabilities add up to 1, as they must (since only the two configurations are possible). [Indeed, we could have used this to calculate $\text{Pr}(E_1)$ without going through the math in the last equation.]

Example T4.1 illustrates one important feature of problems involving Boltzmann factors: *Only the differences between energy levels in the small system have any physical meaning.* One can prove quite generally (see problem T4D.1) that if we add an overall constant to all the energies of the small-system quantum states, we do not change any of the probabilities. This reinforces a principle that we discovered in unit C: physical experiments can *only* define a system's energy up to an overall constant.

Many quantum systems have an infinite number of states, not just two. Calculating Z in such a case can be challenging. But we can still pretty easily calculate relative probabilities, as the next example illustrates.

Example T4.2

Problem: Consider a hydrogen atom on the sun's surface, whose temperature is about 5800 K. What is the probability that such an atom will be in the $n = 2$ energy level, expressed as a fraction of the probability that it is in the ground state ($n = 1$) energy level?

Solution We will assume that the sun is large enough to be considered a reservoir compared to a hydrogen atom (an excellent approximation in this case!). We cannot use $\Pr(E_2) = e^{-E_2/k_B T}/Z$ to calculate the probability here, since in order to calculate $Z = \sum e^{-E_n/k_B T}$ we would have to sum over all states in the hydrogen atom, and there are an infinite number of such states. However, the problem only asks for the probability of the $n = 2$ level *relative* to the $n = 1$ level, so we can use equation T4.5:

$$\frac{\Pr(E_2)}{\Pr(E_1)} = \frac{e^{-E_2/k_B T}}{e^{-E_1/k_B T}} = e^{-(E_2 - E_1)/k_B T} \qquad (T4.16)$$

According to unit Q, the allowed energies E_n for the hydrogen atom are

$$E_n = \frac{E_1}{n^2} \qquad \text{where} \quad E_1 = -13.6 \text{ eV} \quad \text{and} \quad n = 1, 2, 3, \dots \qquad (T4.17)$$

Since we know n for both levels, we can compute the difference $E_2 - E_1$; and since we also know that $T = 5800$ K, we can calculate the ratio. The energy difference in question is

$$E_2 - E_1 = \frac{E_1}{E_2} - E_1 = (-13.6 \text{ eV})\left(\frac{1}{4} - 1\right) = +10.2 \text{ eV} \qquad (T4.18)$$

The value of $k_B T$ here is

$$k_B T = (1.38 \times 10^{-23} \text{ J/K})(5800 \text{ K})\left(\frac{1 \text{ eV}}{1.60 \times 10^{-19} \text{ J}}\right) = 0.50 \text{ eV} \qquad (T4.19)$$

Therefore, the ratio of the probabilities in this case is

$$\frac{\Pr(E_2)}{\Pr(E_1)} = e^{-10.2 \text{ eV}/0.50 \text{ eV}} = e^{-20.4} = 1.38 \times 10^{-9} \qquad (T4.20)$$

Actually, this is the probability of a hydrogen atom's being in any *single* $n = 2$ quantum state. There are actually four distinct quantum states of the hydrogen atom having the same energy E_2 (having different orbital angular momentum components) but only one having energy E_1. So the probability that the atom will be in *any* of the four $n = 2$ levels is actually $4(1.38 \times 10^{-9}) = 5.5 \times 10^{-9}$ times that of its being in the ground state.

So for every billion hydrogen atoms in the ground state on the sun's surface, about 5.5 atoms will be in the first excited state. This is relevant information for astronomers, because only hydrogen atoms already in the $n = 2$ level can absorb *visible* photons, which have enough energy to kick the atoms from the $n = 2$ level into the $n = 3, 4, 5$, and 6 levels. Since these are the *only* transitions that produce hydrogen absorption lines in the visible part of a star's spectrum, the fraction of hydrogen atoms that are ready to make such transitions determines how prominent these absorption lines are.

(If we include the electron's spin orientation in the electron's state, then we actually have *two* distinct $n = 1$ states (corresponding to the two possible orientations of the electron's spin) and *eight* $n = 2$ states. But this does not affect the basic result because including the spin simply multiplies both the top and bottom of the probability ratio by 2.)

T4.3 The Average Energy of a Quantum System

We can also use the Boltzmann factor to calculate the average energy that a quantum system stores when in contact with a reservoir at temperature T.

An analogy may help you see how we can do this. Suppose that during a field biology lab, your lab group is asked to count the number of acorns that squirrels have stored in various dens. The results are shown in table T4.1. Now, to calculate the average number of acorns per den, you *could* type 5 zeros, 20 ones, 47 twos, 18 threes, 7 fours, and 3 fives into your calculator and then divide by 100. If you are a bit more clever and note that adding 20 ones is the same as adding $1 \cdot 20 = 20$, adding 47 twos is the same as adding $2 \cdot 47 = 94$, and so on, you can calculate the average more quickly as follows:

An acorn-counting experiment

Table T4.1   Results of the acorn-counting experiment

Number of Acorns	Number of Dens with This Many Acorns
0	5
1	20
2	47
3	18
4	7
5	3
Total	100

$$\frac{\text{Average acorns}}{\text{Den}} = \frac{0 \cdot 5 + 1 \cdot 20 + 2 \cdot 47 + 3 \cdot 18 + 4 \cdot 7 + 5 \cdot 3}{100} = 2.11 \quad \text{(T4.21)}$$

Now note that 5/100 is the *probability* that a den has zero acorns, 20/100 is the *probability* that a den has one acorn, and so on. Since dividing the sum in the numerator in equation T4.21 by 100 is the same as dividing each term by 100 and then summing, we can also calculate the average in this way:

$$\frac{\text{Average acorns}}{\text{Den}} = 0\left(\frac{5}{100}\right) + 1\left(\frac{20}{100}\right) + 2\left(\frac{47}{100}\right) + 3\left(\frac{18}{100}\right) + 4\left(\frac{7}{100}\right) + 5\left(\frac{3}{100}\right)$$

$$= 0 \cdot \Pr(0) + 1 \cdot \Pr(1) + 2 \cdot \Pr(2) + 3 \cdot \Pr(3) + 4 \cdot \Pr(4) + 5 \cdot \Pr(5) \quad \text{(T4.22)}$$

where $\Pr(n)$ is the probability that a den has n acorns.

Consider now a microscopic quantum system (such as an atom or a molecule) with energy levels E_0, E_1, E_2, and so on. By analogy with the above, we can compute the system's average energy in a given situation by computing

How to find a quantum system's average energy

$$E_{\text{avg}} = E_0 \Pr(E_0) + E_1 \Pr(E_1) + E_2 \Pr(E_2) + \cdots = \sum_{\text{all } n} E_n \Pr(E_n) \quad \text{(T4.23)}$$

where $\Pr(E_n)$ is the probability that the system is in the nth quantum state. But we know from equation T4.8 that if the quantum system is in thermal contact with a reservoir at temperature T, its probability of being in a quantum state with energy E_n is simply

$$\Pr(E_n) = \frac{1}{Z} e^{-E_n/k_B T} \quad \text{where} \quad Z \equiv \sum_{\text{all } n} e^{-E_n/k_B T} \quad \text{(T4.24)}$$

So the average energy of a quantum system in contact with a reservoir is

$$E_{\text{avg}} = \sum E_n \left(\frac{e^{-E_n/k_B T}}{Z}\right) = \frac{\sum E_n e^{-E_n/k_B T}}{\sum e^{-E_n/k_B T}} \quad \text{(T4.25)}$$

The average energy for any quantum system in contact with a reservoir

- **Purpose:** This equation describes how we can calculate the average energy E_{avg} of a quantum system in thermal contact with a reservoir at temperature T, where E_n is the energy of the quantum system's nth quantum state, k_B is Boltzmann's constant, and $Z \equiv \sum e^{-E_n/k_B T}$. The sums are over *all* the system's states.
- **Limitations:** This equation only works for a system in contact with something large enough to be considered a reservoir for the energies the system might typically have.

The average energy of a quantum system (qs) is interesting for a variety of reasons, but one of the most important is the following. Consider a macroscopic object at temperature T that is comprised of N_{qs} identical quantum systems. If N_{qs} is very large, we can imagine each quantum system to be in contact with a reservoir consisting of the remaining systems. Under these circumstances, the object's total thermal energy U is very nearly

$$U = N_{qs}E_{avg} \qquad (T4.26)$$

Therefore, knowing a quantum system's average energy at a given temperature T means that we can calculate the total thermal energy U of an object made up of such quantum systems without having to know anything about the system's multiplicity Ω. Since it is difficult to determine Ω for most systems other than an Einstein solid, this can be *very* useful.

Example T4.3

Problem: Suppose that a certain kind of molecule in a solution at room temperature can be in one of either of two quantum states, which have energies $E_0 = 0$ and $E_1 = \varepsilon$, respectively, where $\varepsilon = 0.02$ eV. What is the average energy of such a molecule? If the solution contains 10^{22} such molecules, what is the total thermal energy U stored by the excited molecules?

Solution The quantum system in this case is an individual molecule; the reservoir is the rest of the gas. Since each molecule is a quantum system, the number of molecules N is the same as the number of quantum systems N_{qs}. In this case, we can write out the sums in equation T4.25 explicitly:

$$E_{avg} = \frac{E_0 e^{-E_0/k_BT} + E_1 e^{-E_1/k_BT}}{e^{-E_0/k_BT} + e^{-E_1/k_BT}} = \frac{0 + \varepsilon e^{-\varepsilon/k_BT}}{e^{-0} + e^{-\varepsilon/k_BT}} = \frac{\varepsilon e^{-\varepsilon/k_BT}}{1 + e^{-\varepsilon/k_BT}} = \varepsilon \frac{1}{e^{\varepsilon/k_BT} + 1} \qquad (T4.27)$$

where in the last step I multiplied top and bottom by e^{ε/k_BT}. At room temperature $k_BT \approx 0.0254$ eV, so $\varepsilon/k_BT = (0.02 \text{ eV})/(0.0254 \text{ eV}) = 0.79$ here. Therefore

$$E_{avg} = \varepsilon \left(\frac{1}{e^{0.79} + 1} \right) = 0.313\varepsilon = 0.00625 \text{ eV} \qquad (T4.28)$$

The total thermal energy U associated with N such molecules is

$$U = NE_{avg} = 0.313N\varepsilon = 0.313(10^{22})(0.02 \text{ eV})\left(\frac{1.60 \times 10^{-19} \text{ J}}{1 \text{ eV}} \right) = 10.0 \text{ J} \qquad (T4.29)$$

Let's check equation T4.28 by using equation T4.8 to calculate the states' probabilities directly and then using equation T4.23 to find the average energy. The probability that the molecule is in the excited state is

$$\Pr(E_1) = \frac{e^{-\varepsilon/k_BT}}{Z} = \frac{e^{-\varepsilon/k_BT}}{e^0 + e^{-\varepsilon/k_BT}} = \frac{1}{e^{\varepsilon/k_BT} + 1} \qquad (T4.30)$$

which is the same as the quantity in parentheses in equation T4.27. So we see in this case that the atom has a probability of 0.313 of being in the excited state and thus a probability of 0.687 of being in the ground state. The average energy is thus $0 \cdot 0.687 + \varepsilon(0.313) = 0.313\varepsilon$.

Now, there is a very cute mathematical trick for evaluating a quantum system's average energy that we will find quite handy as we go along. Let us define $\beta \equiv 1/k_BT$. Then we can rewrite equation T4.25 as follows.

$$E_{avg} = \frac{\sum E_n e^{-E_n \beta}}{\sum e^{-E_n \beta}} \tag{T4.31}$$

Now, note that by the chain rule and the derivative of the exponential

$$\frac{d}{d\beta}(e^{-E_n \beta}) = -E_n e^{-E_n \beta} \tag{T4.32}$$

Because the derivative of a sum is the same as the sum of the derivatives,

$$\frac{dZ}{d\beta} = \frac{d}{d\beta}\left(\sum e^{-E_n \beta}\right) = \sum \frac{d}{d\beta} e^{-E_n \beta} = -\sum E_n e^{-E_n \beta} \tag{T4.33}$$

But this is just the negative of the numerator in equation T4.31. Therefore, we can rewrite that equation as follows:

$$E_{avg} = -\frac{1}{Z}\frac{dZ}{d\beta}, \quad \text{where } Z \equiv \sum e^{-E_n \beta} \text{ and } \beta \equiv \frac{1}{k_B T} \tag{T4.34}$$

- **Purpose:** This equation describes a quick method for calculating a quantum system's average energy E_{avg} if you know its partition function Z as a function of temperature T (k_B is Boltzmann's constant). The sum is over all energy states of the quantum system.
- **Limitations:** This applies only to a quantum system in thermal contact with a reservoir at temperature T.

In this context, calling Z "the partition *function*" makes a *bit* more sense. The energy levels E_n for a given type of quantum system are fixed, so *that* aspect of Z is determined as soon as we decide what type of system we are talking about. But the value of Z *does* depend on T, and thus on β. So in this sense Z is a function of β, and that is the aspect of Z that is interesting here: the *derivative* of this function is related to the system's average energy.

(The "partition" part of the name comes from the fact that Z is instrumental in determining the *probability* of a given state of our quantum system. In a macroscopic system consisting of a large number of identical quantum systems, Z therefore helps specify how many quantum systems are in each energy state, and so helps us "partition" the system into groups of quantum systems having the same state. The Z comes from the German word *Zustandsumme*, meaning "sum over states," which I personally think is a much better description of what Z actually is than the name "partition function.")

As an example, consider again the molecule described in example T4.3. The partition function for this quantum system is

$$Z = e^{-E_0/k_B T} + e^{-E_1/k_B T} = e^{-0} + e^{-\varepsilon/k_B T} = 1 + e^{-\varepsilon \beta} \tag{T4.35}$$

So the average energy of this quantum system is

$$E_{avg} = -\frac{1}{Z}\frac{dZ}{d\beta} = \frac{-1}{1+e^{-\varepsilon\beta}}(0 - \varepsilon e^{-\varepsilon\beta}) = \frac{\varepsilon e^{-\varepsilon\beta}}{1+e^{-\varepsilon\beta}} = \frac{\varepsilon e^{-\varepsilon/k_B T}}{1+e^{-\varepsilon/k_B T}} \tag{T4.36}$$

which is exactly what we found in equation T4.27.

Why Z is called "the partition function"

T4.4 Application to the Einstein Solid

The methods outlined in the last section provide very powerful tools for connecting models of microscopic quantum systems to the observable thermal

Evaluating the average energy
of an oscillator

behavior of macroscopic systems. In this section, I will illustrate this by considering again the Einstein solid.

As we saw in chapter T2, an Einstein solid is essentially a collection of $3N$ independent quantum oscillators. Let's take a single quantum oscillator as our microscopic system, and consider the other $3N - 1$ oscillators in the solid as a reservoir at temperature T (as long as $N \gg 1$, this should be an excellent approximation). The partition function for a single oscillator is

$$Z = e^{-E_0/k_B T} + e^{-E_1/k_B T} + e^{-E_2/k_B T} + \cdots = 1 + e^{-\varepsilon/k_B T} + e^{-2\varepsilon/k_B T} + \cdots$$

$$= 1 + x + x^2 + \cdots \quad \text{where} \quad x \equiv e^{-\varepsilon/k_B T} \tag{T4.37}$$

Note that $x < 1$ in this expression. A well-known mathematical identity tells us that when $x < 1$, this infinite power series has a closed-form equivalent:

$$1 + x + x^2 + \cdots = \frac{1}{1 - x} \tag{T4.38}$$

Also note that

$$\frac{dx}{d\beta} = \frac{d}{d\beta} e^{-\varepsilon\beta} = -\varepsilon e^{-\varepsilon\beta} = -\varepsilon x \tag{T4.39}$$

Therefore, the average energy in a quantum oscillator in an Einstein solid is

$$E_{\text{avg}} = -\frac{1}{Z}\frac{dZ}{dx}\frac{dx}{d\beta} = -(1 - x)\frac{d}{dx}\left(\frac{1}{1-x}\right)(-\varepsilon x) = \frac{\varepsilon x}{1 - x} \tag{T4.40}$$

Exercise T4X.2

Fill in the missing steps in equation T4.40.

This means that the total thermal energy in an Einstein solid with N atoms ($3N$ oscillators) must be

$$U = 3NE_{\text{avg}} = \frac{3N\varepsilon e^{-\varepsilon/k_B T}}{1 - e^{-\varepsilon/k_B T}} = \frac{3N\varepsilon}{e^{\varepsilon/k_B T} - 1} \tag{T4.41}$$

at *any* absolute temperature T. This is much better than we could do in chapter T3, where we could only get a high-temperature limit.

Definition of the "heat
capacity"

Now, we define an object's **heat capacity** C to be dU/dT, the ratio of the amount of heat (or other external energy) dU that we put into the object to the differential temperature change dT that results. (The capital C distinguishes heat capacity from *specific* heat c, which is the heat capacity per unit mass.) C is easier to measure experimentally than an object's total thermal energy U. With a bit of work, we can show that an Einstein solid's heat capacity is

$$C \equiv \frac{dU}{dT} = \frac{d}{dT}\left(\frac{3N\varepsilon}{e^{\varepsilon/k_B T} - 1}\right) = \frac{3Nk_B e^{1/\tau}}{\tau^2(e^{1/\tau} - 1)^2} \quad \text{where} \quad \tau \equiv \frac{k_B T}{\varepsilon} \tag{T4.42}$$

Exercise T4X.3

Fill in the missing steps in equation T4.42.

The Einstein temperature

The quantity τ in this expression is a unitless quantity proportional to the temperature. We can think of it as being a ratio $\tau \equiv T/T_E$ where

$$T_E \equiv \frac{\varepsilon}{k_B} = \text{the solid's \textbf{Einstein temperature}} \tag{T4.43}$$

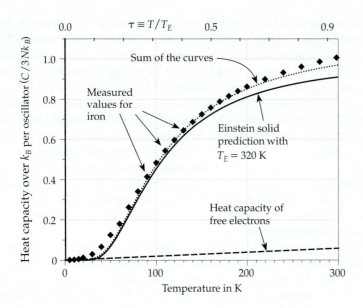

Figure T4.4
The vertical axis shows heat capacity (in units of k_B per oscillator). The solid line shows the prediction of the Einstein model. The diamonds show actual measured results for iron. The dashed line shows the heat capacity contributed by the kinetic energy of electrons that can roam around the metal. The dotted line shows the sum of the two heat capacities. (Data for iron is from Desai, *J. Chem. Ref. Data*, **15**, 3, 1986.)

is a reference temperature that depends on the spacing ε between our particular solid's oscillator energy levels: it is, in fact, the temperature where $k_B T_E = \varepsilon$. This provides a physical temperature scale for the solid.

Comparing with reality

Now equation T4.42 is a pretty complicated function, but fortunately, one can find computer tools (for example, http://www.wolframalpha.com) that make plotting even complicated functions pretty straightforward. The solid line in figure T4.4 shows a graph of this function. Note that at high temperatures, equation T4.42 does imply that $C = dU/dT \approx 3Nk_B$, just as we found in chapter T3. However, the actual data for iron gives a significantly higher value of C near room temperature. This is mostly because the Einstein model does not include the total kinetic energy of the free electrons inside the metal: as the figure shows, including this leads to much better agreement (dotted curve). The data still disagree slightly at the high-temperature end because the potential energy curve for iron atoms is not exactly that of a simple quantum oscillator and the discrepancy becomes more severe as the atoms oscillate more energetically. The significant deviation at the low-temperature end is due to the fact that at low temperatures the idea that the atoms oscillate independently of each other becomes a poor approximation. To address this problem, Peter Debye in 1912 developed a more complicated model (beyond our scope here) that better describes the low-temperature behavior of solids.

However, the Einstein model does pretty well for a simple model. In particular, it describes two crucial features of the experimental curve. The first is that (ignoring the contribution from free electrons), a metal's heat capacity really does level off at $1k_B$ per oscillator ($3k_B$ per atom) at sufficiently high temperatures. This explains the fact (noted in chapter T1) that the heat capacities of *all* monatomic solids are pretty close to $3Nk_B$ at room temperature.

Even more importantly, it predicts (at least qualitatively) the observed sharp decrease in a solid's heat capacity as temperature decreases. We can understand this as follows. At high enough temperatures T, an oscillator's average energy is roughly $k_B T$ (though some oscillators have more and some less). So when the solid has plenty of energy to pass around, each oscillator will tend to get roughly this amount of energy. This is the classical limit of the oscillators' behavior: when each oscillator's energy is high enough that each oscillator contains *many* units of energy ε, we can ignore the fact that a quantum oscillator's energy levels are quantized.

When an oscillator is "frozen out"

But as the solid's temperature T approaches T_E (meaning that $k_B T$ approaches ε), an average oscillator contains just *one* unit of energy. Now the fact that an oscillator's energy levels are quantized has more effect. Note that an oscillator must have at least one unit of energy to have *any energy at all*. As the temperature continues to fall below T_E, the probability that an oscillator has even *one* unit of energy becomes small, and the number of oscillators having exactly *zero* energy begins to weigh disproportionately in the average energy per oscillator, drawing it down below $k_B T$ and the heat capacity per oscillator below k_B. The Boltzmann factor in fact implies that the probability that an oscillator contains even one unit of energy becomes exponentially suppressed as T becomes much smaller than T_E ($k_B T \ll \varepsilon$), so the ability of the solid to store any heat at all also becomes exponentially suppressed. We say that at temperatures $T \ll T_E$, the oscillators are "frozen out" of the normal process of storing energy.

Calculating the inter-atomic "spring constant"

This effect actually provides a means of estimating the effective spring constant of the forces holding an atom in position in a crystal. As discussed in unit Q, the difference ε between a quantum oscillator's energy levels is related to the angular frequency ω at which the oscillator would oscillate classically: $\varepsilon = \hbar\omega$. As we saw in unit N, this angular frequency ω in a classical oscillator is related to the mass of the oscillating object and the spring constant k_s of the spring-like forces that want to hold the object at its equilibrium position: $\omega = (k_s/m)^{1/2}$. We can estimate $T_E = k_B/\varepsilon$ by looking at the curve of a solid's heat capacity (note, for example, that the predicted heat capacity per oscillator is about $0.45k_B$ when $T \approx 0.3T_E$, and this is very close to the observed results, at least for iron), and so compute k_s.

Example T4.4

Problem: What is the effective value of the spring constant k_s for iron atoms?

Solution The solid curve in figure T4.4 was drawn assuming $T_E = 320$ K for iron, but let's estimate T_E assuming we don't know this. The heat capacity per oscillator for iron is about $0.45k_B$ at $T = 100$ K. If this is $0.3T_E$, then T_E is about $(100 \text{ K})/0.3 = 330$ K (which is pretty close to 320 K). Now note that

$$k_s = m\omega^2 = \frac{m(\hbar\omega)^2}{\hbar^2} = \frac{m\varepsilon^2}{\hbar^2} = \frac{m(k_B T_E)^2}{\hbar^2} = m\left(\frac{k_B T_E}{\hbar}\right)^2 \tag{T4.44}$$

So to find k_s, we need the mass of an iron atom. Iron's atomic weight M_A is about 56 g, meaning that Avogadro's number $N_A = 6.02 \times 10^{23}$ iron atoms have this mass. So $m = M_A/N_A$, and

$$k_s \approx \frac{M_A}{N_A}\left[\frac{k_B T_E}{\hbar}\right]^2 = \frac{(0.056 \text{ kg})}{6.02 \times 10^{23}}\left[\frac{(1.38 \times 10^{-23} \text{ J/K})(330 \text{ K})}{1.056 \times 10^{-34} \text{ J·s}}\right]^2\left(\frac{1 \text{ N}}{1 \text{ kg·m/s}^2}\right)$$

$$\approx 170\,\frac{\text{N}}{\text{m}} \tag{T4.45}$$

Note that the units work out: the spring constant expresses the force that a spring exerts when stretched or compressed a certain distance. A macroscopic spring with this spring constant would exert about 1.7 N of force (which is about 0.4 lb) when stretched or compressed by 1 cm.

TWO-MINUTE PROBLEMS

T4T.1   Suppose that an atom has two energy levels separated by an energy difference ΔE. When such atoms are in thermal contact with a reservoir at temperature T, there is roughly 1 atom in the higher state for every 4 in the lower state. What is the approximate ratio of ΔE to $k_B T$?
A. $\frac{1}{4}$
B. 4
C. $\ln 4$
D. $-\ln 4$
E. $e^{-1/4}$
F. e^{-4}
T. Other (specify)

T4T.2   Suppose that an atom has exactly two energy levels separated by an energy difference ΔE. When such atoms are in contact with a reservoir at a temperature T_1, the ratio of the number of atoms in the higher level to the number in the lower level is $\frac{1}{4}$. If we increase the temperature to T_2, the numerical value of this ratio
A. increases.
B. decreases.
C. remains the same.

T4T.3   The quantity Z in the expression $\Pr(E) = Z^{-1} e^{-E/k_B T}$ depends on the temperature T of the reservoir. T or F?

T4T.4   The quantity Z in the expression $\Pr(E) = Z^{-1} e^{-E/k_B T}$ must be
A. always greater than 1.
B. always less than 1.
C. It could be either greater than or less than 1.

T4T.5   As temperatures become large, the denominator in equation T4.41 becomes
A. very large.
B. very small.
C. approximately 1.
D. approximately -1.

T4T.6   Suppose that the energy difference between adjacent energy levels for an oscillator in an Einstein solid is $\varepsilon = 0.005$ eV. At room temperature, the value of T/T_E is
A. quite a bit smaller than 1
B. about equal to 0.5.
C. about equal to 1.
D. about equal to 2.
E. quite a bit larger than 2.
F. One doesn't have enough information to tell.

T4T.7   In the situation described in problem T4T.6, we would expect the average energy of this oscillator when it is in contact with a reservoir at room temperature to be
A. negligible.
B. smaller than $\frac{1}{2}k_B T$.
C. about equal to $\frac{1}{2}k_B T$.
D. about equal to $k_B T$.
E. significantly larger than $k_B T$.
F. One doesn't have enough information to tell.

T4T.8   To calculate an object's heat capacity, we take its specific heat and
A. multiply by the object's mass M.
B. divide by the object's mass M.
C. multiply by the molar mass M_A of the object's molecules.
D. divide by the molar mass M_A of the object's molecules.
E. multiply by the mass m of one molecule in the object.
F. divide by the mass m of one molecule in the object.
T. do something else (specify).

T4T.9   Suppose that the temperature T is about $T_E/9$ for a certain Einstein solid. The probability that an oscillator in that solid will have one unit of energy is closest to
A. 1
B. 1/9
C. 1/81
D. 1/8000
E. $1/10^9$
(*Hint:* See equation T4.38 for how to calculate Z.)

T4T.10 T_E is much smaller for lead than for iron. Which therefore has the greater heat capacity per atom at
(a) $T = 100$ K?
(b) $T = 400$ K?
A. Lead
B. Iron
C. Both should be about the same.

T4T.11   At a sufficiently high temperature, an oscillator in an Einstein solid is more likely to have one unit of energy than no energy at all. T or F?

T4T.12   Suppose that the forces holding an atom in its position in a crystal are the same for two different monatomic solids, but the mass of an atom m is larger for solid B than for solid A. Which has the larger value of T_E?
A. Solid A
B. Solid B
C. Both have about the same value of T_E.

HOMEWORK PROBLEMS

Basic Skills

T4B.1   The value of $k_B T$ at room temperature is roughly $\frac{1}{40}$ eV. (This simple number is worth memorizing, because it helps one tell quickly what kinds of physical processes are possible at room temperature.)
(a) For what temperature T would this be exactly correct?
(b) By about what percent is this off at $T = 295$ K?

T4B.2   Suppose that one atomic oscillator absorbs 100 eV (an enormous amount of energy for a single oscillator) from an Einstein solid containing one thousandth of a mole of atoms (a tiny solid). Assume that the solid's temperature is high enough that its thermal energy is $U \approx 3Nk_B T$.
(a) By roughly how much does the rest of the solid's temperature decrease as the one oscillator absorbs this much energy?
(b) Can we consider the rest of the solid to be a "reservoir" in this case? Justify your response.

T4B.3   The star Vega in the constellation Lyra has a surface temperature of about 9500 K. Estimate the probability that a hydrogen atom on Vega's surface is in a single $n = 2$ energy level as a fraction of the probability of its being in the ground $n = 1$ level.

T4B.4   Suppose that an atom has exactly two energy levels whose energy difference is 0.015 eV. At what temperature will there be two atoms in the higher state for every three in the lower state?

T4B.5   Suppose that an atom has exactly three energy states, with energies 0.020 eV, 0.040 eV, and 0.060 eV, respectively. Out of every 1000 atoms, about how many are in each state at room temperature?

T4B.6   Calculate Z for a single oscillator in an Einstein solid at a temperature $T = 2T_E = 2\varepsilon/k_B$.

T4B.7   Calculate Z for a single oscillator in an Einstein solid at a temperature $T = T_E/2 = \varepsilon/2k_B$.

T4B.8   Suppose you have a system consisting of 10 atoms, and you find that 5 have energy 0 eV, 3 have energy 0.3 eV, 1 has energy 0.8 eV, and 1 has energy 1.5 eV.
(a) Calculate the average energy of an atom in this system by adding up all the energies and dividing by 10.
(b) For each energy value listed above, what is the probability that an arbitrarily chosen atom has that energy?
(c) Use these probabilities to calculate the average energy using equation T4.23.

T4B.9   Suppose that you have an Einstein solid consisting of 5 atoms = 15 oscillators. You find at a given time that six of these oscillators have energy 0 eV, four have energy 0.1 eV, two have 0.2 eV, two have 0.3 eV, and one has 0.4 eV.

(a) Calculate the average energy of an oscillator in this solid by adding up all the oscillators' energies and dividing by 15.
(b) For each energy value listed in this question, what is the probability that an arbitrarily chosen atom has that energy?
(c) Use these probabilities to calculate the average energy using equation T4.23.

T4B.10   Suppose that $Z = k_B T / \varepsilon$ for an atom in a certain substance, where ε is a constant with units of energy. Calculate the average energy per atom in this substance.

T4B.11   Suppose that $Z = ae^{-\varepsilon/k_B T}$ for an atom in a certain substance, where ε is a constant with units of energy and a is a unitless constant. Calculate the average energy per atom in this substance.

Modeling

T4M.1   (Adapted from Schroeder, *Thermal Physics*.) A water molecule can vibrate in many ways, but the "flexing" mode, where the angle between the OH bonds oscillates about its central value of 104°, is the vibrational mode whose excited energies are the smallest (see below).

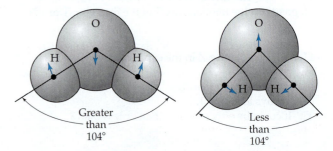

We can model this vibrational mode as if it were a simple one-dimensional quantum harmonic oscillator (mathematically, the situations are equivalent). So as we saw in unit Q, the energy levels are $\frac{1}{2}\hbar\omega$, $\frac{3}{2}\hbar\omega$, $\frac{5}{2}\hbar\omega$, and so on, where ω is the oscillation's angular frequency. This mode's measured value of ω is 3.0×10^{14} s^{-1}. Calculate the absolute probabilities that a water molecule at room temperature will be in its flexing ground state and its first two excited states. (*Hint:* Use equation T4.38 to calculate Z.)

T4M.2   The ground $n = 1$ state of a hydrogen atom is actually two different quantum states with slightly different energies. Because of a magnetic interaction between the proton and electron, the quantum state where the proton and electron spins are aligned has an energy a few microelectronvolts above the energy of the state where the spins are anti-aligned.
(a) If the hydrogen atom undergoes a transition between these states, it emits a photon with a characteristic wavelength of 21 cm that carries away the energy

difference between the states. (Astronomers can detect such photons with a radio telescope and thus determine the distribution of hydrogen gas in the sky: see the photograph below.) The energy of a photon of light with wavelength λ is $E_{ph} = hc/\lambda$. What is the energy difference between the two states?

A map of the 21-cm radio emission from the spiral galaxy M51. (Credit: NASA/CXC/SAO)

(b) If a hydrogen cloud in intergalactic space has a temperature of about 2.9 K (about equal to the temperature of the cosmic background radiation), about what fraction of hydrogen atoms in the $n = 1$ level will be in the aligned state (and thus capable of emitting the 21-cm photons)?

T4M.3 (Adapted from Schroeder, *Thermal Physics*.) At very high temperatures (as could be found in the very early universe), we can model the proton and neutron as being two different quantum states of the same quanton, which we might call a "nucleon." A nucleon can easily be converted from one to the other of its two states by interacting with energetic electrons and/or neutrinos, which were also abundant at the time. The neutron's rest energy is higher than the proton's by about 1.3 MeV (1.3×10^6 eV). At the time when the universe had a temperature of about 10^{11} K, what was the approximate ratio of protons to neutrons?

T4M.4 Below what temperature T are you as likely to find a given oscillator in an Einstein solid to be in the zero-energy state than to be in any other state? Express your answer in terms of T_E. (*Hint:* Use equation T4.38 to calculate Z.)

T4M.5 Suppose that you have a gas whose molecules have two quantum states corresponding to different orientations of a certain subgroup of atoms. The energy difference between these two molecular states is $\Delta E = 0.050$ eV. You are running an experiment where no more than 5% percent of the molecules can be in the higher-energy state, or it will cause noise that will mask the effect you are trying

to measure. Can you run the experiment at room temperature, or do you need to cool your gas?

T4M.6 Suppose that you have a solution containing a substance whose molecule has two quantum states corresponding to different orientations of a certain subgroup of atoms. The energy difference between these two molecular states is $\Delta E = 0.10$ eV. You are running an experiment where no more than 5% percent of the molecules can be in the higher-energy state, or it will cause unacceptable noise. Can you run the experiment at room temperature, or do you need to cool your solution?

T4M.7 An ammonia (NH_3) molecule has an "asymmetric" excited state A and a "symmetric" ground state S. The energy difference between these states is $E_A - E_S = 1.0 \times 10^{-4}$ eV. The ammonia molecule has other possible excited states, but they all have much higher energies relative to the ground state. To what temperature would you need to cool a sample of ammonia molecules so that only $1/3$ of the molecules are in the excited state A?

T4M.8 Suppose that each atom in a certain substance has four quantum states, one with energy 0, two with energy ε, and one with energy 2ε. What is the average energy of an atom in this substance as a function of temperature T? (*Hint:* You can write Z in the form of the square of a binomial: this will make it easier to compute the derivative.)

T4M.9 A quantum mechanical particle moving in one dimension between impenetrable barriers has energy levels ε, 4ε, 9ε, ..., that is, $E_n = \varepsilon n^2$. Suppose that for a certain such quantum system, $\varepsilon = 0.05$ eV. What is the probability that this system will be in its ground state when it is in contact with a reservoir at room temperature? (*Hint:* Drop any terms in Z that are smaller than 10^{-4} times the first term.)

Derivation

T4D.1 Prove that the numerical value of the probability given by equation T4.8 is unchanged if we add a constant value E_0 to the energy of each energy state available to the small system.

T4D.2 Prove equation T4.38. [You can do this a number of ways. The most direct is by long division. If you know that method, divide 1 by $1 - x$ and see what the pattern becomes. Alternatively, you can prove by induction that

$$1 + x + x^2 + \cdots + x^n = \frac{1 - x^{n+1}}{1 - x} \qquad \text{(T4.46)}$$

and then take the limit as n goes to infinity. Yet another way is to expand $(1 - x)^{-1}$ in a Taylor series about $x = 0$, if you know how to do that.]

T4D.3 Show that we can rewrite equation T4.34 in the form

$$E_{avg} = -\frac{d}{d\beta} \ln Z \qquad \text{(T4.47)}$$

T4D.4   When T becomes large compared to $T_E = \varepsilon/k_B$, then $u \equiv \varepsilon/k_B T$ becomes very small and $e^{\varepsilon/k_B T} = e^u$ becomes close to one. Under such circumstances, we can approximate $e^u \approx 1 + u$ (valid when $u \ll 1$). In this limit, show that equation T4.41 becomes $U \approx 3Nk_B T$.

T4D.5   When T becomes large compared to $T_E = \varepsilon/k_B$, then $u \equiv \varepsilon/k_B T = 1/\tau$ becomes very small and $e^{\varepsilon/k_B T} = e^u$ becomes close to one. Under such circumstances, we can approximate $e^u \approx 1 + u$ (valid when $u \ll 1$). In this limit, show that equation T4.42 becomes $dU/dT \approx 3Nk_B$.

T4D.6   When $T \ll T_E$ ($\tau = k_B T/\varepsilon \ll 1$), $1/\tau \gg 1$. What does an Einstein solid's heat capacity dU/dT become in this limit (see equation T4.42)? You should find that this quantity goes to zero (as figure T4.4 shows), but show *how* it goes to zero. That is, what simpler expression could we use to calculate dU/dT when $\tau \ll 1$?

T4D.7   Imagine an object that consists of N particles that each have only two energy states, one with zero energy and the other with energy ε. Find an expression for the heat capacity dU/dT of such an object. (You may use WolframAlpha to calculate the derivative if you specify in your solution what you typed in to get your result.)

T4D.8   (Challenging!) Suppose that a certain kind of molecule has one vibrational quantum state with energy 0, 3 states with energy ε, 5 states with energy 2ε, 7 states with energy 3ε and so on.
(a) Find a closed-form expression for Z in this state. (*Hint:* Express Z in terms of a power series in $x \equiv e^{-\varepsilon/k_B T}$. Look the power series up online or use WolframAlpha to find the closed-form equivalent.)
(b) Find an expression for the vibrational thermal energy U as a function of T for N molecules. (You may use WolframAlpha to calculate the derivative if you specify what you entered to get your result.)
(c) What is the limit of this expression at sufficiently high temperatures (that is, when $T \gg T_E = \varepsilon/k_B$)? (*Hint:* Note that $e^u = 1 + u + u^2/2! + u^3/3! + \cdots$.)
(d) Show that in the low temperature limit ($T \ll T_E$) that $U(T)$ goes exponentially to zero.
(e) Find and plot (as a function of $\tau \equiv T/T_E$) the heat capacity dU/dT of a solid comprised of such molecules. Again, you may use WolframAlpha to calculate the derivative and plot the result if you specify what you entered to get your result. (The derivative is otherwise pretty difficult to do correctly.)

Rich-Context

T4R.1   (Adapted from Schroeder, *Thermal Physics*.) Cold interstellar molecules often contain CN (cyanogen) molecules, whose three first excited rotational quantum states all have an energy of 4.7×10^{-4} eV above that of the single ground rotational state. Studies of absorption spectra of starlight traveling through one cloud done in 1941 implied

that for every 10 CN molecules in the ground state, maybe a bit more than 3 were in the first excited state (that is, an average of a bit more than one in each of the three individual rotational states with the same energy).
(a) To explain these data, astronomers at the time argued that the molecules might be in thermal equilibrium with some "reservoir" at a well-defined temperature. If this is so, what is the approximate temperature T of this reservoir?
(b) In the mid-1960s, physicists discovered that the universe was filled with cosmic background radiation. Anything exposed to this radiation would behave as if it were in thermal contact with a reservoir with a temperature of 2.7 K. Could this be the reservoir that explains these data?

T4R.2   You are doing some research involving a certain diatomic molecule dissolved in a chemical soup of other molecules. This molecule can vibrate along its axis. When the molecule is in its ground vibrational state, it is inert, but in any excited state, it can channel energy to other molecules, instituting a chemical reaction. Your research indicates that if more than 11% of the molecules are in a vibrational excited state, they will initiate a chain reaction that will ruin the experiment. Your research associate claims that this won't happen, because the Boltzmann factor for being in the lowest excited state is 0.10 at room temperature, and the probability of higher excited states is even lower. But another team member notes that there are a lot of possible excited states, and even if each has a low probability, the *total* number of excited molecules could exceed 11%. Which person (if either) is right?

T4R.3   Suppose a certain kind of quantum oscillator is forbidden to be in vibrational states with odd values of n; that is, its energy levels are 0, 2ε, 4ε, 6ε and so on (not ε, 3ε, 5ε, and so on). Find an expression for U as a function of T for N such oscillators. (*Hint:* You can basically appropriate a previous result with very little math if you *think* about it, so your solution to this problem can be simply the final result and an explanation of your reasoning.)

T4R.4   The measured values of the heat capacity of silver divided by $3Nk_B$ for temperatures between 10 K and 150 K are shown in the table below.

T(K)	$\dfrac{1}{3k_B}\dfrac{dU}{dT}$	T(K)	$\dfrac{1}{3k_B}\dfrac{dU}{dT}$
10	0.007	80	0.718
20	0.066	90	0.768
30	0.192	100	0.807
40	0.381	110	0.839
50	0.468	120	0.865
60	0.573	130	0.886
70	0.655	140	0.904

Note that the values approach 1 as the temperature gets higher (indeed, for $T = 295$ K, the value is 1.017). From this data, estimate the effective spring constant of the forces that hold silver atoms to their places in the crystal lattice.

ANSWERS TO EXERCISES

T4X.1   The graph implies that the value of $S_R/(k_B\ln 10)$ drops in a nearly straight line by about 17.5 as the value of U_A increases from 0 to $0.01U = 0$ to 100ε (and the reservoir energy *decreases* by the same amount). Since $\varepsilon = 0.01$ eV, the reservoir's temperature is therefore

$$\frac{1}{T} = \frac{dS_R}{dU_R} = k_B\ln 10\,\frac{d(S_R/k_B\ln 10)}{dU_R} \approx k_B\ln 10\,\frac{-17.5}{-1\text{ eV}}$$

$$\Rightarrow\quad T = \frac{1\text{ eV}}{17.5\,k_B\ln 10} = \frac{1.60\times 10^{-19}\text{ J}}{17.5(1.38\times 10^{-23}\text{ J/K})\ln 10}$$

$$= 288\text{ K} \tag{T4.48}$$

T4X.2   If we define $u = 1 - x$, the chain rule tells us that

$$\frac{d}{dx}\left(\frac{1}{1-x}\right) = \frac{d}{du}\left(\frac{1}{u}\right)\frac{du}{dx} = -\frac{1}{u^2}(-1) = \frac{1}{(1-x)^2} \tag{T4.49}$$

Therefore

$$-(1-x)\frac{d}{dx}\left(\frac{1}{1-x}\right)(-\varepsilon x) = +\frac{(1-x)\varepsilon x}{(1-x)^2} = \frac{\varepsilon x}{1-x} \tag{T4.50}$$

as claimed.

T4X.3   Define $\tau = k_B T/\varepsilon$, $z \equiv 1/\tau$, and $u \equiv e^z - 1$. Then the chain rule tells us that

$$\frac{dU}{dT} = 3N\varepsilon\frac{d}{dT}\left(\frac{1}{e^{\varepsilon/k_B T} - 1}\right) = \frac{d}{du}\left(\frac{1}{u}\right)\frac{du}{dz}\frac{dz}{d\tau}\frac{d\tau}{dT}$$

$$= 3N\varepsilon\left(-\frac{1}{u^2}\right)e^z\left(-\frac{1}{\tau^2}\right)\left(\frac{k_B}{\varepsilon}\right) = \frac{3Nk_B e^{1/\tau}}{\tau^2(e^{1/\tau} - 1)^2} \tag{T4.51}$$

as claimed.

T5 The Ideal Gas

Chapter Overview

Introduction

In this chapter, we deploy the techniques developed in the last chapter to explore the thermal behavior of gases.

Section T5.1: Quantum Particles in a One-Dimensional Box

A gas consists of particles bouncing around a mostly empty container. The first step toward modeling this is to consider a particle moving in one dimension between two barriers a distance L apart. Quantum mechanics tells us that such a particle's x-momentum will be quantized: $p_x = \pm hn/2L$, where h is Planck's constant and n is a positive integer. This means that the particle's energy is also quantized: $E_n = \varepsilon n^2$, where $\varepsilon \equiv h^2/8mL^2$. For a typical container and any reasonable temperature T, ε is so small compared to $k_B T$ that we can approximate the sum $Z = \sum e^{-\varepsilon n^2/k_B T}$ for the partition function by an integral. Once we know the partition function as a function of $\beta \equiv 1/k_B T$, we can calculate the average energy of such a particle. The result is

$$E_{avg} = -\frac{1}{Z}\frac{dZ}{d\beta} = \frac{1}{2}k_B T \tag{T5.7}$$

Section T5.2: A Three-Dimensional Monatomic Gas

We can make the model more realistic by extending it to three dimensions. In such a case, the particle's momentum components p_x, p_y, and p_z are separately and independently quantized. We can separate the particle's energy $E = |\vec{p}|^2/2m = (p_y^2 + p_y^2 + p_z^2)/2m$ into three terms, each of which is independent of the other.

The section argues that whenever a quantum system has energy storage modes A and B that are separate and independent, the partition function for the whole is the product of the partition functions for the independent modes: $Z = Z_A Z_B$. Moreover, we can calculate the average energies for each mode *separately* from its particular partition function and simply add the final results to get the total energy.

In the particular case of a three-dimensional monatomic gas, the total kinetic energy of an average molecule at sufficiently high temperature is the sum of contributions from independent partition functions for the x, y, and z directions:

$$K_{avg} = E_{avg} = \frac{3}{2}k_B T \tag{T5.11}$$

This means that the heat capacity of a monatomic gas should be $dU/dT = \frac{3}{2}Nk_B$, which is consistent with empirical results.

Section T5.3: Diatomic Gases

The difference between monatomic gases and other gases is that multi-atom molecules can store energy in modes *in addition* to their kinetic energy (the kinetic energy is $K_{avg} = \frac{3}{2}k_B T$ for *all* gases). In particular, diatomic molecules can store energy in the form of rotational energy. The energy levels for this mode are $E_j = j(j+1)\varepsilon$ (where $\varepsilon \equiv h^2/8\pi^2 I$, h is Planck's constant, and I is the molecule's moment of inertia), and there are $2j + 1$ quantum states per energy level. In the high-temperature limit, we can approximate the partition function's sum as an integral. Taking the derivative of the result, we find that $E_{rot,avg} = k_B T$. This means that the total heat capacity of a diatomic gas is $dU/dT = \frac{5}{2}Nk_B$, which is again close to what we observe.

Single atoms are not solid entities that can rotate, so this energy storage mode is not available to atoms in monatomic gases.

Section T5.4: The Equipartition Theorem and Its Limits

Note that the average energy per atom in an Einstein solid, the average kinetic energy of a gas molecule, and the average rotational energy of a diatomic molecule are all integer multiples of $\frac{1}{2}k_B T$. The equipartition theorem provides a non-quantum approach to understanding this result.

The equipartition theorem states that if the Newtonian expression for the energy of a molecule contains a term that depends on the square of some classical variable q (such as x-position, x-momentum p_x, or the x component L_x of angular momentum), then the average energy associated with that term is $\frac{1}{2}k_B T$. The corresponding variable q is called a **degree of freedom**. The section illustrates how the theorem yields the high-temperature results we have derived earlier. In particular, both the equipartition theorem and our earlier quantum calculations agree that

$$U \approx \frac{f}{2}Nk_B T, \quad \frac{dU}{dT} \approx \frac{f}{2}Nk_B \qquad \text{(T5.21)}$$

- **Purpose:** These equations specify the thermal energy U and the heat capacity dU/dT of simple gases containing N particles at temperature T, where f is the gas molecule's "degrees of freedom" ($f = 3$ for monatomic gases, $f = 5$ for simple diatomic gases) and k_B is Boltzmann's constant.
- **Limitations:** These expressions are approximations valid at sufficiently high temperatures. They also assume that vibrational modes of molecular energy storage either do not exist or are not significant.

However, only the quantum results yield the correct low-temperature behavior. When the temperature falls so low that even the first excited energy level in an energy storage mode is pretty improbable, then that mode has difficulty storing any energy at all and the heat capacity associated with that level goes to zero compared to its high-temperature limit. We say that the mode in such a case is "frozen out."

Section T5.5: The Ideal Gas Law

At normal temperatures, a Newtonian model adequately expresses the movement of gas atoms. Such a model implies that gas molecules moving with an average kinetic energy of $\frac{3}{2}k_B T$ bouncing off of the gas container's walls will exert a **pressure** (force per unit area) on those walls such that

$$PV = Nk_B T \qquad \text{(T5.28)}$$

- **Purpose:** This equation gives the pressure P of N molecules of a low-density gas held in a volume V at temperature T, where k_B is Boltzmann's constant.
- **Limitations:** This equation strictly applies only in the zero-density limit (although it is a good approximation for real gases at typical densities). All parts of a gas sample must be in equilibrium for it to have well-defined single values of P and T throughout the volume.
- **Notes:** This is the **ideal gas law:** any gas obeying this law is an **ideal gas**.

Each species in a mixture of gases exerts a partial pressure $P_s = N_s k_B T/V$, and the total pressure is the sum of partial pressures. Also the chemist's ideal gas law is $PV = nRT$, where n is the number of moles and $R = N_A k_B = 8.31$ J/K is the **gas constant**.

T5.1   Quantum Particles in a One-Dimensional Box

Gases are extremely important systems in thermal physics. We started our exploration of thermal physics with the Einstein solid because one can quite easily determine a macrostate's multiplicity for this particular kind of object. We cannot so easily do this for gases. However, the techniques developed in the last chapter will enable us to do a clever end-run around this problem.

What is a gas?

What is the difference between a gas and a solid? Molecules in a solid are bound together in such a way that a given molecule vibrates around a fixed position in a lattice. The distance between molecules is also basically the same as the size of a molecule. But gas molecules are *not* bound to each other; instead, they may freely roam around whatever container holds the gas. The average separation of gas molecules is typically much larger than the molecule's diameter, meaning that a gas consists mostly of empty space. This means that a gas's volume is not essentially fixed (as a solid's volume is); indeed, specifying a gas's volume V is an important part of specifying its macrostate.

An idealized one-dimensional gas model

Let's begin with the simplest possible quantum model of such a situation. Suppose that we have N gas particles that are moving completely freely back and forth in a *one-dimensional* container between two impenetrable barriers a distance L apart. Also assume that the particles simply pass through each other without interacting as they move back and forth, so that each particle behaves as if it were *alone* in the container, interacting only with the barriers. This is a "spherical cow" type of model, but will get us started.

Let the x axis of our reference frame be the axis along which the particles move. Quantum mechanics tells us that in the situation described above, each particle's x-momentum magnitude $|p_x|$ is *quantized*:

$$|p_x| = \left(\frac{h}{2L}\right)n \quad \text{where } n = 1, 2, 3, \dots \tag{T5.1}$$

and h is Planck's constant. Unit Q discusses why: in this unit we will accept this as given (see problem T5D.1 if you are interested).

If our gas particle is a completely structureless mathematical point, the only energy it can have is kinetic energy. Assuming that the particle is non-relativistic and has mass m, this energy is

$$E = K = \tfrac{1}{2}mv_x^2 = \frac{(mv_x)^2}{2m} = \frac{p_x^2}{2m} = \frac{1}{2m}\left(\frac{hn}{2L}\right)^2 = \frac{h^2n^2}{8mL^2} \tag{T5.2}$$

Note that because $|p_x|$ is quantized, the energy is also quantized. This means that the partition function for a single particle is

$$Z = \sum_{n=1}^{\infty} e^{-\varepsilon n^2/k_B T} \quad \text{where } \varepsilon = \frac{h^2}{8mL^2} \tag{T5.3}$$

where ε is a characteristic energy for this system and the sum is over all possible quantum states.

Approximating the partition function using an integral

Unlike the oscillator case, this infinite sum has no easy closed-form equivalent. However, we can *approximate* this sum pretty easily, as follows. Suppose we draw a picture where we represent the values of $e^{-\varepsilon n^2/k_B T}$ at a given temperature T but at different values of n as a sequence of bars of width $\Delta n = 1$ (see figure T5.1). If $\varepsilon \ll k_B T$, so that the exponential decays very slowly as n increases, then the value of the sum is approximately equal to the value of the area under the curve of the exponential function from 0 to infinity. Therefore, we can approximate the sum as an integral:

$$Z = \sum_{n=1}^{\infty} e^{-\varepsilon n^2/k_B T}\,\Delta n \approx \int_0^{\infty} e^{-\varepsilon n^2/k_B T}\,dn \quad \text{if } \varepsilon \ll k_B T \tag{T5.4a}$$

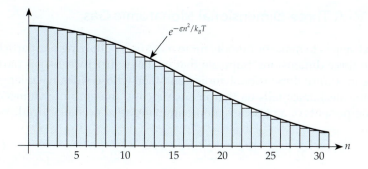

Figure T5.1

The area under the curve of the function $e^{-\varepsilon n^2/k_B T}$ is approximately equal to the area of the bars, which is equivalent to the sum in the partition function. Note that if ε were even smaller, the curve would become flatter, and the approximation improves (the neglected areas between the bars and curve become even smaller).

This is actually a superb approximation at room temperature. You can show that for a helium atom ($m \approx 6.7 \times 10^{-27}$ kg) trapped between barriers only a micron apart ($L = 10^{-6}$ m), then $\varepsilon/k_B T \approx 2 \times 10^{-9}$ at room temperature. This means that n^2 must be $\sim 10^8$ ($n \sim 10^4$) for $e^{-\varepsilon n^2/k_B T}$ to fall significantly below 1 even in this extreme case. The exponential therefore decreases *very* slowly with increasing n indeed (the actual curve is *much* flatter than the one shown in figure T5.1).

Exercise T5X.1

Verify that $\varepsilon/k_B T \approx 2 \times 10^{-9}$ for helium in a micron-sized container at room temperature. Also argue that $\varepsilon/k_B T$ will be even smaller if m and/or L are larger (note that helium is the lightest possible monatomic gas).

Now note that in the integral, we are treating n not as an integer but as a continuous variable. We can put it into a more evocative form if we rename n to u (a variable without integer connotations) and define $a^2 \equiv \varepsilon/k_B T$: then

$$Z \approx \int_0^\infty e^{-a^2 u^2}\,du \qquad \text{where} \quad a \equiv \sqrt{\frac{\varepsilon}{k_B T}} \tag{T5.4b}$$

We call the function $e^{-a^2 u^2}$ a **Gaussian**: it is the classic "bell-shaped curve." The integral of a Gaussian is not very easy to calculate from scratch (see problem T5D.2 for a discussion), but it is easy to look up online. The result is

$$Z \approx \left[\frac{\sqrt{\pi}}{2a}\right] = \sqrt{\frac{\pi k_B T}{4\varepsilon}} = \sqrt{\frac{\pi}{4\varepsilon\beta}} \qquad \text{where} \quad \beta \equiv \frac{1}{k_B T} \tag{T5.5}$$

We finally are in a position to be able to evaluate the average energy of a gas particle when it is in thermal contact with the reservoir comprised of all of the rest of the gas particles:

The average energy of a particle in a one-dimensional gas

$$E_{\text{avg}} = -\frac{1}{Z}\frac{dZ}{d\beta} \approx -\sqrt{\frac{4\varepsilon\beta}{\pi}}\frac{d}{d\beta}\sqrt{\frac{\pi}{4\varepsilon\beta}} = -\sqrt{\frac{4\varepsilon\beta}{\pi}}\sqrt{\frac{\pi}{4\varepsilon}}\frac{d}{d\beta}\beta^{-1/2}$$

$$= \beta^{1/2}\left(-\frac{1}{2}\beta^{-3/2}\right) = \frac{1}{2\beta} = \frac{k_B T}{2} \tag{T5.6}$$

Therefore, the total thermal energy of our one-dimensional gas is

$$U = NE_{\text{avg}} = \tfrac{1}{2}Nk_B T \tag{T5.7}$$

We have therefore successfully determined the gas's total thermal energy without having to calculate the gas's multiplicity or entropy!

T5.2   A Three-Dimensional Monatomic Gas

Generalizing to three dimensions

The next step is to make our model more realistic by allowing our particles to move in three dimensions. Suppose that we have structureless gas particles moving in a *three*-dimensional cubic container whose sides have length L. Quantum mechanics tells us that in such a case, the absolute values of all three components of a particle's momentum are *separately* quantized:

$$|p_x| = \left(\frac{h}{2L}\right)n_x, \quad |p_y| = \left(\frac{h}{2L}\right)n_y, \quad |p_z| = \left(\frac{h}{2L}\right)n_z \qquad (T5.8)$$

where n_x, n_y, and n_z are positive and *completely independent* integers. The particle's total energy is

$$E = \frac{p_x^2 + p_y^2 + p_z^2}{2m} = \frac{h^2}{8mL^2}(n_x^2 + n_y^2 + n_z^2) = \varepsilon n_x^2 + \varepsilon n_y^2 + \varepsilon n_z^2 \qquad (T5.9)$$

where (as before) $\varepsilon \equiv h^2/8mL^2$. The particle's total partition function will be

$$Z = \sum_{n_x=1}^{\infty} \sum_{n_y=1}^{\infty} \sum_{n_z=1}^{\infty} e^{-E/k_BT} = \sum_{n_x=1}^{\infty} \sum_{n_y=1}^{\infty} \sum_{n_z=1}^{\infty} e^{-\varepsilon(n_x^2 + n_y^2 + n_z^2)/k_BT} \qquad (T5.10a)$$

This looks really horrible to calculate. But note that $e^{-(a+b+c)} = e^{-a}e^{-b}e^{-c}$. This means that we can pull out the exponential factor that depends only on n_x in front of the sums over n_y and n_z, and we can pull out the exponential factor that depends only on n_y in front of the sum over n_z, and so do each sum separately:

$$Z = \sum_{n_x=1}^{\infty} e^{-\varepsilon n_x^2/k_BT} \left(\sum_{n_y=1}^{\infty} e^{-\varepsilon n_y^2/k_BT} \left(\sum_{n_z=1}^{\infty} e^{-\varepsilon n_z^2/k_BT}\right)\right)$$

$$= \sum_{n_x=1}^{\infty} e^{-\varepsilon n_x^2/k_BT} \left(Z_z \sum_{n_y=1}^{\infty} e^{-\varepsilon n_y^2/k_BT}\right) = Z_y Z_z \sum_{n_x=1}^{\infty} e^{-\varepsilon n_x^2/k_BT} = Z_x Z_y Z_z$$

$$\text{where} \quad Z_x \equiv \sum_{n_x=1}^{\infty} e^{-\varepsilon n_x^2/k_BT} \quad \text{(and similarly for } y \text{ and } z) \qquad (T5.10b)$$

Now we can evaluate each Z factor separately just as we did in equations T5.4 and T5.5: they each should be $Z_x = Z_y = Z_z = \sqrt{\pi/4\varepsilon\beta}$, meaning that the partition function and average energy for a given particle in the gas is

$$Z = Z_x Z_y Z_z = \left(\frac{\pi}{4\varepsilon\beta}\right)^{3/2} \quad \text{and} \quad E_{\text{avg}} = -\frac{1}{Z}\frac{dZ}{d\beta} = \frac{3}{2}k_BT \qquad (T5.11)$$

Exercise T5X.2

Show that $E_{\text{avg}} = \frac{3}{2}k_BT$ follows from $Z = (\pi/4\varepsilon\beta)^{3/2}$.

We see that the average energy is just the *sum* of the average energies $\frac{1}{2}k_BT$ connected with each of the particle's possible one-dimensional motions.

We can handle partition functions for independent variables separately

This illustrates a very important general principle. If a quantum system has two completely *independent* ways A and B of storing energy, then its total partition function is simply the *product* of the partition functions associated with energy-storage mode A and energy-storage mode B: $Z = Z_A Z_B$. When we calculate the system's average energy, we find that

$$E_{\text{avg}} = -\frac{1}{Z}\frac{dZ}{d\beta} = -\frac{1}{Z_A Z_B}\frac{d}{d\beta}(Z_A Z_B) = -\frac{1}{Z_A Z_B}\left(\frac{dZ_A}{d\beta}Z_B + Z_A\frac{dZ_B}{d\beta}\right)$$

$$= -\frac{1}{Z_A}\frac{dZ_A}{d\beta} - \frac{1}{Z_B}\frac{dZ_B}{d\beta} = E_{\text{avg},A} + E_{\text{avg},B} \qquad (\text{T5.12}a)$$

Note also that if we calculate the Boltzmann factor for a given quantum state of mode A, while summing over all states in mode B (meaning that we don't care what state in mode B the system has) we find that

$$\Pr(E_{nA}) = \frac{1}{Z}\sum_{\text{all }j} e^{-(E_{nA}+E_{jB})/k_B T} = \frac{e^{-E_{nA}}}{Z_A}\frac{1}{Z_B}\sum_{\text{all }j} e^{-E_{jB}/k_B T} = \frac{e^{-E_{nA}}\, \cancel{Z_B}}{Z_A\, \cancel{Z_B}} \qquad (\text{T5.12}b)$$

These very important results say that

> We can analyze each *independent* mode of energy storage in a quantum system *separately*, ignoring any other storage modes, and simply add the results to find the total energy.

This is the ultimate justification for the examples in the previous chapter where we focused on just one aspect of a quantum system's energy (for example, a hydrogen atom's internal energy level in a stellar atmosphere) while ignoring other aspects of the atom's energy (such as its kinetic energy).

Now let's turn back to the specific case of the three-dimensional gas. What if all sides of the gas's container did not have the same length? This would mean that ε would have different values for each of the three directions. But you can easily show that these distinct values of ε still cancel out of the calculation for E_{avg} (see problem T5D.3), leaving the result unchanged. Though it is much more difficult to show, the container doesn't even have to be rectangular: its shape is completely irrelevant.

The container's shape is not relevant

So we see from equation T5.11 that the total thermal energy U and heat capacity dU/dT of non-interacting gas particles in an arbitrary container are

The heat capacity of a monatomic gas

$$U = NE_{\text{avg}} = \frac{3}{2}Nk_B T \qquad \Rightarrow \qquad \frac{dU}{dT} = \frac{3}{2}Nk_B \qquad (\text{T5.13})$$

As noted in chapter T1, this is *exactly what we observe* experimentally for *monatomic* gases (such as helium, argon, krypton, and so on). The reason that this still-simplistic model works well for these noble gases is that their atoms really do not interact significantly, so that model assumption is sound. For reasons that we will discuss in the next section, such atoms have no other accessible ways of storing energy, so the atoms of a noble gas really do behave pretty much as if they were non-interacting structureless particles!

T5.3 Diatomic Gases

The difference between monatomic gases and other gases is that multi-atom molecules can store energy in modes *other* than their kinetic energy. Because these modes are almost always independent of the molecules' center-of-mass kinetic energy, the results of the previous section ensure that a molecule's average *kinetic* energy is $K_{\text{avg}} = \frac{3}{2}k_B T$ for *all* gases. But for multi-atom molecules, this is no longer the entire energy.

The kinetic energy of any kind of gas molecule is still $\frac{3}{2}k_B T$

Diatomic gas molecules, for example, can store internal energy in the form of rotational energy. Quantum mechanics tells us that if a molecule has a moment of inertia I, its rotational energy levels are

$$E_j = j(j+1)\varepsilon \quad \text{where } j = 0, 1, 2, \dots \text{ and } \quad \varepsilon = \frac{h^2}{8\pi^2 I} \qquad (\text{T5.14})$$

There are also $2j + 1$ distinct quantum states corresponding to each energy level E_j. (The derivation of these results is a topic for an upper-level quantum physics class, so we'll simply accept them here as given.) The partition function for this energy storage mode is therefore

$$Z_{\text{rot}} = \sum_{j=0}^{\infty} (2j + 1) e^{-j(j+1)\varepsilon/k_B T} \tag{T5.15}$$

To evaluate this sum, we apply the same trick that we did in section T5.1: as long as $\varepsilon \gg k_B T$, then we can approximate the sum by an integral:

$$Z_{\text{rot}} = \sum_{j=0}^{\infty} (2j + 1) e^{-j(j+1)\varepsilon/k_B T} \Delta j \approx \int_0^{\infty} (2j + 1)\, e^{-j(j+1)\varepsilon/k_B T}\, dj \tag{T5.16}$$

where $\Delta j = 1$ in the sum, but j in the integral is no longer an integer but a continuous variable. Even for diatomic molecules with fairly small moments of inertia I, the value of $\varepsilon/k_B \approx$ a few kelvins, so at room temperature, $\varepsilon/k_B T \approx 100$. Therefore, the exponential still varies pretty slowly and the integral is a good approximation to the sum (though not as good as it was in section T5.1). Most diatomic gases (except for hydrogen) become liquids before the temperature would be low enough so that this is much of an issue.

This integral *looks* much harder than the Gaussian integral we considered before, but it is actually quite easy with a clever substitution (see problem T5D.6). In any case, the result is simply

$$Z_{\text{rot}} \approx \frac{k_B T}{\varepsilon} = \frac{1}{\varepsilon\beta} \tag{T5.17}$$

The average energy per rotating molecule

The average energy that a diatomic molecule stores in this mode is therefore

$$E_{\text{avg,rot}} = -\frac{1}{Z_{\text{rot}}} \frac{dZ_{\text{rot}}}{d\beta} \approx -\varepsilon\beta \frac{d}{d\beta}\left(\frac{1}{\varepsilon\beta}\right) = -\varepsilon\beta\left(\frac{-1}{\varepsilon\beta^2}\right) = \frac{1}{\beta} = k_B T \tag{T5.18}$$

The total thermal energy of a diatomic gas in the high-temperature limit is the sum of the molecules' kinetic and rotational energies:

$$U = NE_{\text{avg}} = N(E_{\text{avg,KE}} + E_{\text{avg,rot}}) = N(\tfrac{3}{2}k_B T + k_B T) = \tfrac{5}{2}Nk_B T \tag{T5.19a}$$

So the heat capacity of a diatomic gas should be

$$\frac{dU}{dT} = \frac{5}{2}Nk_B T \tag{T5.19b}$$

Again, as noted in chapter T1, this is pretty close to what we observe. The main problem is that gas molecules, especially those more complicated than the simplest diatomic molecules, can also store energy in the form of various kinds of flexing vibrations. Therefore, the heat capacities of all but the simplest diatomic molecules tend to be somewhat higher than $\frac{5}{2}Nk_B T$.

Why single atoms cannot store rotational energy

Note that a single atom *cannot* store rotational energy in this way. A single atom is simply an electron cloud surrounding a tiny nucleus, and the electron cloud is not a solid thing that can rotate as a unit like a ball can. We can increase the angular momenta of the cloud's individual electrons, but this involves moving the electrons to energy levels so high that they are simply not accessible at normal temperatures.

T5.4   The Equipartition Theorem and Its Limits

Now, we have so far studied three different ways that an atom or molecule can store energy (vibrational energy in an Einstein solid, kinetic energy in a gas, rotational energy in a gas), and in each case, the average amount of energy stored in each mode has been an integer multiple of $\frac{1}{2}k_B T$.

The **equipartition theorem** provides a simple way to remember this. This theorem states that if you can write a system's *Newtonian* energy as a sum of terms involving squares of independent position or momentum components, then *each term* gets an average of $\frac{1}{2}k_B T$ of energy. For example, we would say that the Newtonian energy of an atom in a monatomic gas is

Describing the equipartition theorem

$$E = K = \frac{|\vec{p}|^2}{2m} = \frac{p_x^2}{2m} + \frac{p_y^2}{2m} + \frac{p_z^2}{2m} \qquad (T5.20a)$$

where p_x, p_y, and p_z are its momentum components and m is its mass. By the equipartition theorem, each of this equation's three independent terms gets $\frac{1}{2}k_B T$ of energy on the average, meaning that the atom's total average energy in a monatomic gas is $\frac{3}{2}k_B T$, as we found in the last section. The energy of an atom of mass m oscillating in one dimension in an Einstein solid is given by

$$E = K + V = \frac{p_x^2}{2m} + \frac{1}{2}k_s x^2 \qquad (T5.20b)$$

where k_s is the effective spring constant. By the equipartition theorem, each of the terms in this equation gets $\frac{1}{2}k_B T$ of energy on the average, meaning that the atom's total average energy while oscillating in one dimension is $k_B T$, exactly as we found in chapter T4. The Newtonian energy of a diatomic gas molecule with mass m and moment of inertia I is

$$E = K + U^{\text{rot}} = \frac{|\vec{p}|^2}{2m} + \frac{|\vec{L}|^2}{2I} = \frac{p_x^2}{2m} + \frac{p_y^2}{2m} + \frac{p_z^2}{2m} + \frac{L_x^2}{2I} + \frac{L_y^2}{2I} \qquad (T5.20c)$$

where L_x and L_y are components of the molecule's angular momentum around axes perpendicular to the molecule's long axis (taken to be the z axis here). L_z is not included because a diatomic molecule cannot rotate around that long axis (for pretty much the same reasons that a single atom cannot rotate around any axis). Since the energy equation has five independent terms (or **degrees of freedom** in the theorem's language), each term gets $\frac{1}{2}k_B T$ of energy on the average, and the molecule's total average energy is $\frac{5}{2}k_B T$.

The equipartition theorem suggests that a **polyatomic** molecule (a molecule with more than two atoms, such as water, ammonia, or methane) should have an extra degree of freedom because rotations around all three axes are permitted. However, the situation with such molecules is complicated in practice because such molecules can also often store energy in the form of flexing vibrational modes. Therefore, polyatomic molecules do *not* reliably store the $3k_B T$ of average energy that the equipartition theorem implies.

In summary, to appropriate the language of the equipartition theorem, we have seen that both the theorem and the quantum mechanical calculations presented in the last two sections agree that for simple monatomic and diatomic gases at room temperature,

$$U \approx \frac{f}{2}Nk_B T, \quad \frac{dU}{dT} \approx \frac{f}{2}Nk_B \qquad (T5.21)$$

- **Purpose:** These equations specify the thermal energy U and the heat capacity dU/dT of simple gases containing N particles at temperature T, where f is the gas molecule's "degrees of freedom" ($f = 3$ for monatomic gases, $f = 5$ for simple diatomic gases) and k_B is Boltzmann's constant.
- **Limitations:** These expressions are approximations valid at sufficiently high temperatures. They also assume that vibrational modes of molecular energy storage either do not exist or are not significant.

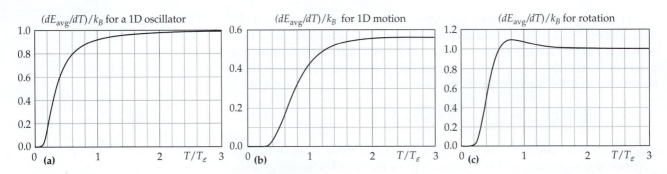

Figure T5.2

The predicted heat capacity per storage mode for (a) a one-dimensional quantum oscillator, (b) a quantum particle moving in one dimension between two barriers, and (c) a rotating diatomic molecule.

Low-temperature behavior

The equipartition theorem's roots precede quantum mechanics, and has some justification in pre-quantum statistical mechanics (see problem T5D.4 for a discussion of the reasoning behind it). But the theorem has been superceded by the quantum calculations that I have already presented. In addition to providing a more rigorous way of counting states, only quantum calculations can provide an accurate understanding of what happens at low temperatures. When temperatures are low enough that the integral approximation falls apart, we can in the quantum case return to actually doing the sums directly. This is pretty easy to do with a computer. Figure T5.2 shows computer-generated graphs of the heat capacity of a one-dimensional oscillator, a particle moving in one dimension, and a rotating diatomic molecule (respectively), all plotted as a function of T/T_ε, where $T_\varepsilon \equiv \varepsilon/k_B$.

Note that these functions have some intriguing differences. The heat capacity of a particle moving in one dimension actually overshoots $\frac{1}{2}k_B$; though off the right end of the graph, it curves *very slowly* back down toward $\frac{1}{2}k_B$. It is still about $0.51k_B$ even at $T/T_\varepsilon = 1000(!)$, but $T/T_\varepsilon \sim 10^9$ (typical for gases at room temperature), the function has solidly returned to being $\frac{1}{2}k_B$. The rotational heat capacity also overshoots k_B, but much more quickly returns to that value.

However, all of these curves share an important feature: at sufficiently *low* temperatures, the heat capacities drop sharply and approach zero with zero slope. As discussed in the last chapter, this is because at temperatures such that $k_B T \ll \varepsilon$, even the lowest quantum energy level above the zero-energy level becomes very improbable, meaning that the system cannot store significant amounts of energy at all.

If a quantum system's energy levels were *not* quantized, then it would *always* behave like a Newtonian system (which can have any energy) and therefore would always have the heat capacity predicted by the equipartition theorem. But because energies are quantized, the system deviates significantly from that expected behavior. We say that at sufficiently low temperatures, degrees of freedom can be "frozen out" by these quantum effects.

Implications for the hydrogen gas molecule

We can see these effects in operation in the heat capacity of hydrogen (see figure T5.3). Because hydrogen atoms are so light, a hydrogen molecule's moment of inertia is much lower than any other diatomic molecule, so T_ε is roughly 350 K instead of being a few kelvins (as it is for gases like nitrogen and oxygen). Therefore, this mode of energy storage doesn't fully settle in until about $T = 400$ K. (Some subtle quantum issues erase the "bump" seen in figure T5.2.) Below about 60 K, this energy storage mode is completely frozen out, so between 20 K (where hydrogen liquifies) and 60 K, hydrogen behaves like a monatomic gas, not a diatomic gas!

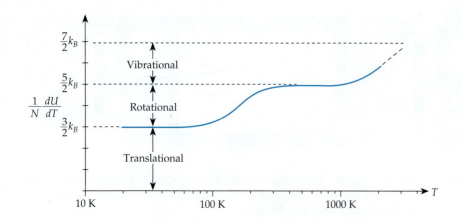

Figure T5.3
A (mostly qualitative) graph of the heat capacity per molecule of hydrogen gas at constant volume and standard pressure as a function of absolute temperature. (Adapted from Schroeder, *Thermal Physics*, Addison-Wesley, 2000, p. 30.)

Above 1000 K, the temperature is high enough that the hydrogen molecule can begin to store energy by vibrating along its axis, but this mode is only partially switched on when hydrogen begins to disassociate at 2000 K.

T5.5   The Ideal Gas Law

We saw in the previous section that the average *kinetic* energy per molecule in any gas must be $\frac{3}{2}k_B T$. We can use this to predict the pressure that the gas molecules exert on a wall of the gas's container.

Because the differences between the molecular kinetic energy levels are incredibly tiny compared to a molecule's average energy, quantum effects are negligible and we can safely treat the molecules as Newtonian particles. The simplest Newtonian gas model assumes that

A Newtonian gas model

1. The gas consists of a huge number of identical particles of mass m.
2. These particles have negligible volume compared to their separation.
3. They don't interact with each other.
4. They rebound elastically from collisions with the container walls.
5. They move at constant velocities between collisions.

We'll see how to relax some of these restrictions shortly.

We define the **pressure** that the gas exerts on its container's walls to be the *force per unit area* exerted on those walls. The SI unit of pressure is the pascal (1 Pa = 1 N/m^2): a gas whose pressure is comparable to that of the earth's atmosphere exerts about 10^5 Pa on its container walls. This pressure is due to molecules endlessly bouncing off the container's walls: each collision with a container wall exerts a tiny outward push on that wall. Because the molecules are moving randomly in all directions (without any direction being preferred), the pressure is the same on *any* surface exposed to the gas.

The definition of *pressure*

To more easily quantify the *magnitude* of the push that these collisions exert, suppose that we confine a sample of gas in a cylinder by a piston that is free to move frictionlessly in response to molecules hitting it. By the definition of force, the total average force that the gas exerts on the piston is the average rate at which these collisions transfer momentum to the piston:

Quantifying the pressure exerted on a wall by colliding gas molecules

$$\left[\vec{F}\right]_{\text{avg}} \equiv \left[\frac{d\vec{p}}{dt}\right]_{\text{avg}} \tag{T5.22}$$

We can keep the piston from moving in response to this force by exerting an opposing force of equal magnitude; in what follows, assume that we do this.

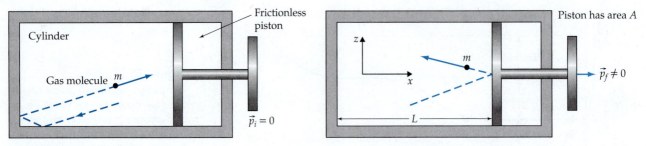

Figure T5.4
When a molecule hits an otherwise isolated piston, the molecule transfers an impulse to the piston. The total force that the gas exerts on the piston is the rate at which such collisions transfer momentum to the piston. (In practice, we will exert an external force on the piston to hold it at rest against the force exerted by the gas.)

The pressure exerted by a single molecule

For the sake of simplicity, let's focus on a *single* gas molecule of mass m bouncing around the chamber (see figure T5.4). Each time it hits the piston, we assume that it rebounds elastically, which means that kinetic energy is conserved in the collision. Since we are holding the piston at rest, conservation of kinetic energy implies that the *molecule's* kinetic energy, and so its speed, remains the same. The collision in fact simply reverses the component of the molecule's velocity perpendicular to the piston's face (v_x in the case shown in figure T5.4). The change in the molecule's x-momentum is thus

$$p_x \text{ (final)} - p_x \text{ (initial)} = -m|v_x| - m|v_x| = -2m|v_x| \tag{T5.23}$$

Conservation of momentum implies that such a collision delivers an *opposite* x-impulse $[dp_x] = +2m|v_x|$ to the piston.

Now, every time the molecule collides with a cylinder wall, either its x-velocity is unaffected (as when the molecule hits an upper or lower wall) or it is reversed without changing magnitude (as when the molecule hits the piston or the opposite wall). So the value of $|v_x|$ is *not* changed by a collision with any wall, and because the molecule moves freely between collisions, $|v_x|$ *never* changes in this model. Because the molecule must cover a distance of $2L$ (the length of the cylinder and back) between collisions with the piston, the time interval between collisions of this particular molecule with the piston is $2L/|v_x|$. Therefore, the magnitude of the average force that this particular molecule exerts on the piston is

$$[F]_{\text{avg}} = |F_x|_{\text{avg}} = \frac{\text{impulse per collision}}{\text{time between collisions}} = \frac{2m|v_x|}{2L/|v_x|} = \frac{mv_x^2}{L} \tag{T5.24}$$

The pressure P is the force per unit area, so

$$P \text{ (for one molecule)} \equiv \frac{[F]_{\text{avg}}}{A} = \frac{mv_x^2}{AL} = \frac{mv_x^2}{V} \tag{T5.25}$$

since $AL = V$ is the cylinder's volume.

The pressure exerted by N identical molecules

Now, the pressure exerted by N non-interacting molecules is just the sum of the pressures due to each molecule. But the molecules do not necessarily move at the same speed as they rattle around the cylinder (we will see how their speeds are distributed in the next chapter). So we cannot simply multiply equation T5.25 by N: we must actually do the sum. Let P_i be the pressure due to the ith molecule, $v_{i,x}$ be that molecule's x-velocity, and m_i be its mass. Then (by definition of the average)

$$P \text{ (total)} = \sum_{i=1}^{N} P_i = \sum_{i=1}^{N} \frac{m_i v_{i,x}^2}{V} = \frac{N}{V}\left(\frac{1}{N}\sum_{i=1}^{N} m_i v_{i,x}^2\right) \equiv \frac{N}{V}[mv_x^2]_{\text{avg}} \tag{T5.26}$$

Now, if the gas is at rest, the molecules will be moving in random directions, so we should have $[v_x^2]_{\text{avg}} = [v_y^2]_{\text{avg}} = [v_z^2]_{\text{avg}}$ (no coordinate direction should be different than any other coordinate direction). Therefore

$$[v^2]_{\text{avg}} = [v_x^2 + v_y^2 + v_z^2]_{\text{avg}} = [v_x^2]_{\text{avg}} + [v_y^2]_{\text{avg}} + [v_z^2]_{\text{avg}} = 3[v_x^2]_{\text{avg}}$$

$$\Rightarrow \quad [mv_x^2]_{\text{avg}} = \tfrac{1}{3}[mv^2]_{\text{avg}} = \tfrac{1}{3}[2K]_{\text{avg}} = \tfrac{1}{3}(2\tfrac{3}{2}k_B T) = k_B T \qquad \text{(T5.27)}$$

where K is a molecule's kinetic energy. So our simple model predicts that the pressure exerted by any gas should be $P = (N/V)k_B T$, or, equivalently,

$$PV = Nk_B T \qquad \text{(T5.28)}$$

The ideal gas law

- **Purpose:** This equation specifies the pressure P of N molecules of a low-density gas enclosed in a volume V at temperature T, where k_B is Boltzmann's constant.
- **Limitations:** This equation strictly applies only in the zero-density limit (although it is a good approximation for real gases at typical densities). All parts of a gas sample must be in equilibrium for it to have well-defined single values of P and T throughout the volume.
- **Notes:** We call this equation the **ideal gas law,** and any gas obeying this law is an **ideal gas.**

We can think of the ideal gas law as describing the general behavior of gases in the abstract, behavior that is perturbed by the specific characteristics of a given real gas. The main problems with our model is that gas molecules do not have zero size relative to their separation and that they do interact with each other. However, as the gas's density decreases, both approximations become better. The densities of gases at atmospheric pressure and room temperature are generally low enough that deviations from the ideal gas law are tolerably small (although measurable).

Chemists usually write the ideal gas law in the form

$$PV = nRT \qquad \text{(T5.29a)}$$

The chemist's form of the ideal gas law

where (using $N_A \equiv$ Avogadro's number $= 6.02 \times 10^{23}$ molecules per mole) $n \equiv N/N_A =$ the number of moles of the gas in question and

$$R \equiv N_A k_B = \left(\frac{6.02 \times 10^{23}}{1 \text{ mol}}\right)\left(\frac{1.38 \times 10^{-23} \text{ J}}{1 \text{ K}}\right) = \frac{8.31 \text{ J}}{1 \text{ mol} \cdot \text{K}} \qquad \text{(T5.29b)}$$

(The constant R is often called **the gas constant.**) A modest advantage of this form is that we can easily determine n from macroscopic measurements

$$n = \frac{M}{M_A} \qquad \text{(T5.30)}$$

where M is the mass of the sample and M_A is the gas's **molar mass** (sometimes incorrectly called its *molar weight*), defined to be the mass of a mole (that is, Avogadro's number of molecules) of the gas, typically expressed in units of grams per mole. In turn, M_A can be quickly calculated by summing the atomic masses (which one can find listed in any periodic table) of all atoms appearing in the gas molecule. The value of M_A in grams per mole is also *approximately* the total number of nucleons (protons and neutrons) in the molecule's atoms. For example, a nitrogen molecule contains 2 nitrogen atoms, each with 7 protons and 7 neutrons, so the molecular mass of nitrogen is approximately $2 \times 14 = 28$ g/mol.

In this unit, though, we want to explain gas properties in terms of the *microscopic* behaviors of its molecules, so the physicist's version of the law (with its explicit reference to the number of molecules N) will be generally more meaningful, and I will use it consistently throughout the unit. Note that it is still pretty easy to calculate N for a given sample of gas:

$$N = N_A n = \frac{N_A}{M_A} M \qquad (T5.31)$$

Do not confuse n (the number of moles) with N (the number of molecules).

Exercise T5X.3

Suppose you have a bottle containing 1400 cm^3 of N_2 at atmospheric pressure 101 kPa ($1 \text{ Pa} = 1 \text{ N/m}^2$) and room temperature 295 K. How many molecules are in the bottle? What is the mass of the gas?

Mixtures of gases

Since the gas molecules in the ideal gas model do not interact, we can consider gases that contain a mixture of chemical species as being completely separate gases that happen to occupy the same volume. The ideal gas law applies to each species separately

$$P_s V = N_s k_B T \qquad (T5.32)$$

where T is the common temperature of the mixture, N_s is the number of molecules of species s in the common volume V, and P_s is the **partial pressure** that the molecules of that species s exert on the container walls. Note that we can calculate N_s for each species by using equation T5.31 and the molecular mass $M_{A,s}$ for that species. Just as the total number of molecules in the mixture is the sum of N_s over all species, the total pressure exerted by the mixture is just the sum of the partial pressures:

$$P = \sum_{\text{all species}} P_s \qquad (T5.33)$$

Making the ideal gas model more realistic

Now, assuming that the gas molecules do not collide with each other and bounce only elastically off the container walls made our analysis simpler, but it is not realistic. Relaxing these assumptions, however, does not actually change the results. Collisions between molecules will redirect those molecules in random directions and modify their energies, but as long as their sizes are negligible compared to their separation, they will still (on the average) arrive at a given container wall as often as they would if they were traveling unimpeded, because the *average* kinetic energy and thus the *average* value of v_x^2 is constant. Similarly, a collision between a gas molecule and a molecule on the container wall will lead to random energy transfers and generally redirects the gas molecule's velocity in a random direction, but *on the average*, those collisions will behave as if they were elastic, because in equilibrium no *net* energy is transferred between the wall and the gas molecules.

The main inadequacy of the ideal gas model is that the finite size of molecules means that the volume through which they can freely travel is less than the container's volume by the total volume of the gas molecules. This means that in dense gases the molecules hit the container walls a bit more often than expected, meaning that their pressure is a bit higher than expected. Any long-range interactions that exist between molecules can also make the molecular speeds slightly different than expected. But these effects are typically small at normal gas densities.

TWO-MINUTE EXERCISES

T5T.1 Which of the partition functions below yields an average energy E_{avg} that is NOT an integer multiple of $\frac{1}{2}k_B T$ in the high-temperature (small-β) limit? Assume that ε is a constant with units of energy, and $\beta = 1/k_B T$.
A. $Z = (2\pi/\varepsilon\beta)^2$
B. $Z = (\varepsilon\beta)^{-1/2}$
C. $Z = (1 - e^{-\varepsilon\beta})^{-1}$ (note that $e^x \approx 1 + x$ when $|x| \ll 1$)
D. None of the above.
E. All of the above.

T5T.2 Using an integral to calculate a sum over $f(n)$ (where n is an integer) is a good approximation when
A. the summed values $f(n)$ are individually small.
B. the value of $f(n)$ varies only very slowly with n.
C. the value of $f(n)$ goes to zero as n goes to infinity.
D. the value of $f(n)$ goes to zero as n goes to zero.
E. All of the conditions above are true.

T5T.3 Which of the following energy storage modes for a gas molecule "switch on" fully at
(a) the lowest temperature?
(b) the highest temperature?
A. The molecule's kinetic energy
B. The molecule's rotational energy
C. The molecule's vibrational energy

T5T.4 The heat capacity of any energy storage mode with quantized energy levels goes to zero as the temperature T goes to zero. T or F?

T5T.5 An atom of helium can store energy by bumping an electron from its lowest orbital energy state to a higher orbital energy level. In particular, moving an electron from the lowest state to the next-lowest state would store an energy of 24.6 eV. Why can we ignore this energy storage mode when calculating the heat capacity of helium gas?
A. This mode is "frozen" out at normal temperatures.
B. Collisions between atoms can't influence the energy levels of electrons *inside* the atoms.
C. This storage mode is completely independent of the kinetic and/or rotational energy modes.
D. Only modes involving helium *molecules* count.
E. We have ignored this mode only for simplicity's sake.
F. Some other reason (specify).

T5T.6 **(a)** The average x-velocity of molecules in a container of gas at rest is zero. T or F? **(b)** The average molecular speed is zero. T or F?

T5T.7 If the speed of a molecule in a container doubles (other things remaining the same), the average pressure that this molecule exerts on the container wall
A. remains the same.
B. doubles.
C. quadruples.
D. One does not have enough information to say.

T5T.8 Suppose that when we add 20.8 J to 1 mol of a certain gas, its temperature is observed to increase by 1 K. Which of the following statements is true?
A. This gas is monatomic.
B. This gas is diatomic.
C. This gas is polyatomic.
D. One cannot determine the type of gas without information that we are not given.
E. The results described are impossible.

T5T.9 The ideal gas model assumes that molecules have infinitesimal size. Consider *one* molecule bouncing around in the container. As the molecule's size increases relative to the size of the container, how will the average pressure P that it exerts change, other things being held constant?
A. P increases.
B. P does not change.
C. P decreases.
D. One does not have enough information to say.

T5T.10 Two identical rooms, A and B, are sealed except for an open doorway between them. Room A is warmer than B. Which room has the greater number of molecules? (*Hint:* How will the pressures in the two rooms compare?)
A. Room A
B. Room B
C. Both have the same number of molecules.

T5T.11 One mole of hydrogen gas and $\frac{1}{2}$ mole of helium gas are mixed in a container and maintained at a fixed temperature T. How does the total pressure P_H exerted by the hydrogen molecules compare to the total pressure P_{He} exerted by the helium molecules? Note that a molecule (atom) of helium has twice the mass of an H_2 molecule.
A. $P_H = 2P_{He}$
B. $P_H = \sqrt{2}P_{He}$
C. $P_H = P_{He}$
D. $P_H = \sqrt{\frac{1}{2}}P_{He}$
E. $P_H = \frac{1}{2}P_{He}$
F. Other (specify)

T5T.12 N molecules of a monatomic gas A are mixed with the same number of molecules of a diatomic gas B in a container, and the mixture is held at a constant temperature T (which is about equal to room temperature). The molecules of gas A have mass m_A, and the molecules of gas B have mass $m_B > m_A$. Which gas has
(a) the greater value of $[v^2]_{avg}$,
(b) the greater average kinetic energy per molecule,
(c) the greater total thermal energy U, and
(d) the greater pressure P?
In each case, choose from one of the four answers below.
A. The quantity is greater for gas A.
B. The quantity is greater for gas B.
C. The quantity is the same for both gases.
D. One does not know enough to compare the quantities.

HOMEWORK PROBLEMS

Basic Skills

T5B.1   Does the partition function for an arbitrary molecular energy storage mode become larger or smaller as the temperature increases? Explain your reasoning.

T5B.2   What is $\varepsilon/k_B T$ at room temperature for the kinetic energy storage mode for nitrogen gas in a cubic container whose sides have length $L = 20$ cm?

T5B.3   What is $\varepsilon/k_B T$ at room temperature for the kinetic energy storage mode for hydrogen gas in a cubic container whose sides have length $L = 50$ cm?

T5B.4   An atom of helium can store energy by bumping its electron from its lowest orbital energy state to a higher orbital energy level. In particular, moving an electron from the lowest state to the next-lowest state would store an energy of 24.6 eV. Explain why we can ignore this energy storage mode when calculating the heat capacity of helium gas at ordinary temperatures.

T5B.5   Calculate the approximate temperatures at which the following energy storage modes of a nitrogen molecule become "unfrozen" (that is, when their heat capacities become at least half their respective maximum values).
(a) The kinetic energy mode ($\varepsilon \sim 10^{-7}$ eV)
(b) The rotational energy mode ($\varepsilon \sim 0.00025$ eV)
(c) The vibrational energy mode ($\varepsilon \sim 0.29$ eV)

T5B.6   Assuming that the bond lengths are the same, for which gas would the rotational energy storage mode "freeze out" at the lower temperature, hydrogen gas or deuterium gas? (Deuterium is like hydrogen except that its nucleus has a neutron in addition to the proton.) Explain.

T5B.7   (a) What is the total thermal energy of 0.40 g of helium gas inside a balloon at room temperature? (b) How fast would you have to run to have that much kinetic energy?

T5B.8   Interstellar space has about one H_2 molecule per cubic centimeter. The temperature of deep space is about 3 K.
(a) What is the pressure of this interstellar gas?
(b) How does this compare to the pressure of the best vacuum we can achieve in the laboratory ($\approx 10^{-13}$ Pa)?

T5B.9   (a) Estimate the number of air molecules in your dorm room or bedroom. (You will have to estimate the room's size.) (b) Does it matter what kinds of molecules the air in your room actually contains? Why or why not?

T5B.10   Dry air consists of 78% nitrogen, 21% oxygen, and 1% argon by weight. Show that the mass of 1 mol of air is about 29.0 g.

T5B.11   Consider a gas whose molecules have an average kinetic energy of 1 eV. What is the temperature of the gas?

T5B.12   The rms ("root-mean-square") speed of gas molecules is $v_{rms} \equiv \left([v^2]_{avg}\right)^{1/2}$. Find v_{rms} of CO_2 molecules in a room at room temperature (295 K). (Carbon atoms have 12 and oxygen atoms have 16 nucleons in their nuclei.)

T5B.13   The rms ("root-mean-square") speed of gas molecules is $v_{rms} \equiv \left([v^2]_{avg}\right)^{1/2}$. If v_{rms} of Cl_2 molecules in a sample of gas is about 180 m/s, what is the temperature of the gas? (Chlorine atoms have 35 nucleons in their nuclei.)

T5B.14   The average *velocity* of a molecule in a gas at rest is zero, but the average *speed* of such a molecule is not zero. Explain how this is possible.

Modeling

T5M.1   Consider a diatomic gas at temperatures T_1 and T_2 such that $\tau_1 \equiv k_B T_1/\varepsilon = 0.29$ and $\tau_2 = 0.31$ for the gas's rotational energy storage mode.
(a) Evaluate the partition function directly as a sum for each temperature, keeping what you think is a sufficient number of terms needed to ensure six significant figures of accuracy (justify your choice).
(b) For T_1 and T_2, compute $E_{avg} = \sum \Pr(E_j)E_j$ (in terms of ε) directly as a sum with the same number of terms.
(c) Note that $dE_{avg}/d\tau \approx [E_{avg}(\tau_2) - E_{avg}(\tau_1)]/[\tau_2 - \tau_1]$ as long as $\tau_2 - \tau_1$ is sufficiently small. Use this approximation to estimate the value of $(dE_{avg}/dT)/k_B$ at $\tau = 0.30$.
(d) Compare your result with that shown in figure T5.2c. Are you close? (You should be!)

T5M.2   HD gas molecules have one hydrogen atom bound to one deuterium atom (a deuterium atom is a hydrogen atom with a neutron as well as a proton in its nucleus). Suppose that we find that at 26 K, the heat capacity per molecule is almost exactly $2k_B$. What is the rotational moment of inertia of this molecule? (*Hint:* Consult figure T5.2c to determine ε.)

T5M.3   A water molecule's flexing oscillation mode (see problem T4M.1) has a measured oscillation angular frequency of $\omega = 3.0 \times 10^{14}$ Hz. Note that a simple oscillator's energy levels are separated by $\hbar\omega$, where $\hbar = h/2\pi$ and h is Planck's constant. Is this energy storage mode mostly "switched on" or "switched off" at room temperature?

T5M.4   One can currently use nanotechnology to construct wire segments small enough to trap an electron so that it must move in one dimension between two effectively impenetrable barriers $L = 7.0$ nm apart.
(a) Calculate ε for an electron in this situation.
(b) At room temperature, how much of the substance's heat capacity (per electron) is due to the trapped electron's kinetic energy? (*Hint:* Use figure T5.2b.)

T5M.5   Suppose that an insulating container of air at room temperature and standard pressure is initially traveling at a speed $|\vec{v}_0|$. We bring the container suddenly to rest. Assume that the air's center-of-mass kinetic energy gets converted entirely to thermal energy in this process.

(a) What is the air's change in temperature ΔT in terms of k_B, $|\vec{v}_0|$, and the average mass of an air molecule m?

(b) If $|\vec{v}_0| = 30$ m/s (about highway speed), what is the numerical value of ΔT? (*Hint:* See problem T5B.10.)

T5M.6 Suppose that, during a hailstorm, hailstones with an average mass of 1 g falling vertically with an average speed of 10 m/s hit the roof of your car at a rate of 300 hits per second. Assume that they bounce elastically from the roof, and that your roof is about 2 m long and 1.5 m wide.

(a) What total force do the hailstones exert on the roof?

(b) What is the pressure that they exert on the roof?

(c) How does this pressure compare to air pressure?

T5M.7 Suppose that we add 250 J of energy to 1000 cm^3 of air initially at room temperature and normal pressure. By about how much does its pressure increase if its volume remains constant?

T5M.8 Suppose that we add 100 J of energy to 1000 cm^3 of helium gas initially at room temperature and normal pressure. By about how much does its volume increase if its pressure remains constant?

Derivations

T5D.1 In quantum mechanics, a free particle is represented by a wavelike "wavefunction" whose wavelength is connected to its momentum by the de Broglie relation $|\vec{p}| = h/\lambda$, where h is Planck's constant. When a particle must move in one dimension along the x axis, then only its x-momentum p_x is nonzero. If the particle is trapped between two impenetrable barriers, then its wavefunction must go to zero at each boundary.

(a) Argue that this means that n half-wavelengths of the particle's wavefunction must fit within the distance L between the two boundaries, where n is an integer.

(b) Show that this means that $|p_x| = hn/2L$.

(In Newtonian mechanics, a particle with a given energy would bounce back and forth between the barriers, spending equal time moving in each direction. The quantum state likewise embraces both possible directions even-handedly. This is why the state determines $|p_x|$ but does not determine the sign of p_x.)

T5D.2 (Requires knowing some multivariable calculus.) Here is a clever trick for calculating the integral $\int e^{-a^2 x^2}\, dx$, where a is a constant. Note that

$$\left(\int_0^\infty e^{-a^2 x^2}\, dx \right)\left(\int_0^\infty e^{-a^2 y^2}\, dy \right) = \int_0^\infty \int_0^\infty e^{-a^2 x^2} e^{-a^2 y^2}\, dx\, dy$$

$$= \int_0^\infty \int_0^\infty e^{-a^2(x^2 + y^2)}\, dx\, dy = \int_0^\infty \int_0^\infty e^{-a^2 r^2}\, dx\, dy \quad \text{(T5.34)}$$

where $r^2 \equiv x^2 + y^2$. But we can convert this integral to polar coordinates r and ϕ, where $dx\, dy = r\, dr\, d\phi$:

$$\int_0^\infty \int_0^\infty e^{-a^2 r^2}\, dx\, dy = \int_0^\infty \int_0^{\pi/2} e^{-a^2 r^2} r\, dr\, d\phi \quad \text{(T5.35)}$$

(a) Explain why the limits have changed.

(b) Both the r and ϕ integrals are pretty easy to do in this case. Do the integral in equation T5.35. (*Hint:* Use substitution of variables: $u \equiv a^2 r^2$.)

(c) Show, therefore, that

$$\int_0^\infty e^{-a^2 x^2}\, dx = \frac{\sqrt{\pi}}{2a} \quad \text{(T5.36)}$$

T5D.3 Suppose that $Z_x = \sqrt{\pi/4\varepsilon_x \beta}$, $Z_y = \sqrt{\pi/4\varepsilon_y \beta}$, and $Z_z = \sqrt{\pi/4\varepsilon_z \beta}$, where ε_x, ε_y, and ε_z are different values (because the distances between the barriers in the x, y, and z directions are different). Show that the value of E_{avg} corresponding to $Z = Z_x Z_y Z_z$ is still $\frac{3}{2} k_B T$.

T5D.4 (Adapted from Schroeder, *Thermal Physics*, Addison-Wesley, 2000.) In this problem, I will outline the pre-quantum argument for the equipartition theorem. Consider a storage mode whose Newtonian energy is given by $E(q) = cq^2$, where q is some variable and c is an arbitrary constant. Though q is a continuous variable in Newtonian mechanics, we will "count" states by *pretending* that q has discrete, evenly spaced values separated by some very tiny step Δq. If Δq is sufficiently tiny, we can approximate the sum that yields the partition function by an integral: defining $\beta = 1/k_B T$, we have

$$Z = \sum_{\text{all states}} e^{-cq^2 \beta} = \frac{1}{\Delta q} \sum_q e^{-cq^2 \beta}\, \Delta q \approx \frac{1}{\Delta q} \int_{-\infty}^\infty e^{-cq^2 \beta}\, dq \quad \text{(T5.37)}$$

(a) Evaluate Z. (*Hint:* See equation T5.5. Note that the integral from $-\infty$ to 0 will be the same as that from 0 to ∞.)

(b) Show that $E_{avg} = \frac{1}{2} k_B T$, no matter what c and Δq are. Now the variable q (which could be position x, x-momentum p_x, angular momentum x component L_x, the y or z components of the same quantities, or something similar) is actually continuous in Newtonian mechanics. We can restore q's continuity by taking the limit as Δq goes to zero. This would cause the value of Z to become infinite, but we see that the value of E_{avg} does not change in this limit.

T5D.5 (Adapted from Schroeder, *Thermal Physics*, Addison-Wesley, 2000.) Consider an energy storage mode whose energy is given by $E(q) = c|q|$, where q is some continuous Newtonian variable and c is a constant. (This formula describes the kinetic energy of a one-dimensional relativistic gas.) Adapt the argument outlined in problem T5D.4 to this situation and show that the average energy of this storage mode is $E_{avg} = k_B T$.

T5D.6 Evaluate the integral

$$\int_0^\infty (2j + 1) e^{-j(j+1)\varepsilon/k_B T}\, dj \quad \text{(T5.38)}$$

(*Hint:* Change variables to $u = j(j+1)\varepsilon/k_B T$.)

T5D.7 Suppose that the energy levels of a certain energy storage mode in a quantum system are given by $E_n = \varepsilon n^2$ and there are n^2 quantum states per energy level.

(a) Look up the integral you need to evaluate Z in the high-temperature limit.

(b) Determine E_{avg} in this limit.

T5D.8   Suppose the energy levels of a certain energy storage mode in a quantum system are given by $E_n = \varepsilon n$, and there are n quantum states per energy level.

(a) Use the fact that

$$\sum_{n=1}^{\infty} nx^n = \frac{x}{(1-x)^2} \qquad (T5.39)$$

to evaluate the sum for Z exactly.

(b) Evaluate E_{avg} for this energy storage mode.

(c) Evaluate E_{avg} in the high-temperature (low-β) limit. (*Hint:* Note that $e^z \approx 1 + z$ when $z \ll 1$.)

T5D.9   Consider a cylindrical parcel of air of area A and infinitesimal height dz. If this air parcel is to remain stationary, the difference between the total pressure forces exerted on its top and bottom faces must be equal to its weight.

(a) Use this information and the ideal gas law to show that

$$\frac{dP}{dz} = -\frac{m|\vec{g}|}{k_B T} P \qquad (T5.40)$$

where m is the mass of an air molecule, $|\vec{g}|$ is the gravitational field strength, k_B is Boltzmann's constant, T is the absolute temperature, and P is the air pressure at the vertical position z of the parcel's center.

(b) Assuming that T is independent of height in the earth's atmosphere (a pretty crude approximation), find a formula for the air pressure P as a function of altitude z.

(c) Use this to estimate the air pressure at the top of Mount Everest at $z = 8848$ m above sea level as a fraction of the pressure at sea level (assuming $T \approx 295$ K).

T5D.10   Consider the partial pressure law (equation T5.33).

(a) Show that this law follows from the ideal gas law.

(b) Use the Newtonian particle model to explain qualitatively why this *physically* makes sense.

T5D.11   *Boyle's law* (first stated by Robert Boyle in 1662) claims that the pressure of a given sample of gas at a given temperature is inversely proportional to its volume. *Gay-Lussac's law* (first stated by Joseph Louis Gay-Lussac in 1802) claims that a gas's volume is proportional to its absolute temperature at constant pressure. Amedeo Avogadro argued in 1811 that, at constant pressure and temperature, the volume of a sample of gas is proportional to the number of molecules it contains, independent of the type of gas. Prove that all three laws follow from the ideal gas law.

Rich-Context

T5R.1   The magnitude of upward buoyant force on a balloon is equal to the magnitude of weight of the air that it displaces. Assume that the mass of the unfilled balloon and its payload is M.

(a) Argue that if the buoyant force is approximately in balance with the gravitational force on the balloon (so that the balloon floats), we must have

$$\rho_o - \rho_i = \frac{M}{V} \qquad (T5.41)$$

Figure T5.5
A hot air balloon. Note the people in the basket: this will help you estimate the balloon's size. Credit: © McGraw-Hill Education

where ρ_o is the density of the air outside the balloon, ρ_i is the density of the gas inside the balloon, and V is the balloon's volume.

(b) Argue that the density of an ideal gas with molar mass M_A, pressure P, and absolute temperature T is

$$\rho = \frac{M_A P}{RT} \qquad (T5.42)$$

where $R = N_A k_B = 8.31$ J/K.

(c) In a typical hot-air balloon (see figure T5.5), the bottom of the balloon is open to the surrounding atmosphere, so the pressure of the hot air inside the balloon must be essentially equal to that of the cold air surrounding the balloon. If the air inside a typical hot-air balloon is roughly at 100°C = 373 K, and the surrounding air is at about 17°C = 290 K, about how much volume must the balloon have? (*Hint:* See problem T5B.10.) What is the radius of a sphere with the same volume? Does your answer make sense, considering the picture shown in figure T5.5?

T5R.2   One of the hazards of space travel is the possibility that a meteoroid might punch a hole in your space vehicle, allowing the air inside to leak away into the vacuum of space. Roughly how long would you have to live if a meteoroid punched a hole with an area $a \approx 1$ cm^2 in your spacecraft? The answer to this question may be interesting not only to you but also to the *designers* of the spacecraft. For example, providing hole-patching materials is pointless if the astronauts will only live for a few seconds.

To make the problem more concrete, assume we are talking about a reasonably sized module of the International Space Station. Assume that this module is sealed off from the rest of the other modules; it is shaped like a cylinder roughly 4 m in diameter and 10 m long; and the hole has been punched in one of the end faces of the cylinder.

(a) About how often does a given molecule hit the end face of the module? What is the approximate probability that if it does, it will go through the hole? Combine the answers to these questions to argue that the number of molecules of air that escape through a hole of area a per unit time in a module of volume V is given as

$$\left|\frac{dN}{dT}\right| \approx \frac{a|v_x|_{avg}}{2V}N \tag{T5.43}$$

where N is the number of molecules in the module, v_x is the x component of a given molecule's velocity, and the average is over all molecules. Note that the average here is of $|v_x|$, not v_x (why not?).

(b) Show that this equation implies that $N(t) = N_0 e^{-t/\tau}$, where N_0 is the number of molecules present at time $t = 0$ and $\tau \equiv 2V/a|v_x|_{avg}$.

(c) Given this, how long do you think someone can survive before blacking out (in terms of τ)? Explain how you are making your estimate. (*Hint:* At the top of Mt. Everest, the air pressure is about $1/3$ that at sea level.)

(d) It is challenging to calculate $|v_x|_{avg}$ exactly, but argue that $|v_x|_{avg} \approx \sqrt{k_B T/m}$, where m is the mass of a molecule. Why is this an approximation?

(e) Put these results together to answer the question. About how much time do you have to patch the hole?

T5R.3 Consider two identical imaginary boxes of air with volume V, one at sea level (altitude $z = 0$) and one at an altitude z that is much larger than the size of either box. A molecule of mass m in a given quantum state in the upper box will have an energy that is $m|\vec{g}|z$ higher than the exactly corresponding quantum state in the lower box because of the extra gravitational potential energy involved with being in the upper box. Now, suppose that we connect these boxes with a hose so that molecules are free to flow between the boxes. Suppose also that the gases in both boxes have the same absolute temperature T.

(a) Argue that the Boltzmann factor implies that, in equilibrium, the ratio of the number $N(z)$ of molecules in the upper box to the number $N(0)$ in the lower box must be $N(z)/N(0) = \exp(-m|\vec{g}|z/k_B T)$, where $\exp(x) \equiv e^x$.

(b) Argue that the ratio of gas *pressures* must be

$$\frac{P(z)}{P(0)} = \exp(-m|\vec{g}|z/k_B T) \tag{T5.44}$$

(c) We can model the earth's atmosphere as a series of vertically stacked and connected boxes of gas, so the last equation should apply to the earth's atmosphere as long as the entire atmosphere has a temperature independent of z (this is the isothermal atmosphere model). Use this model to predict the approximate air pressure at the top of Mount Everest ($z = 8848$ m) if $T \approx 295$ K.

(d) What is the approximate *density* of air at Coors Stadium in Denver, Colorado ($z = 1610$ m)?

(e) Argue that this significantly affects how far a well-hit baseball travels in that stadium compared to baseball stadiums at coastal cities. (*Hint:* Consult unit N.)

Comment: The earth's atmosphere is *not* isothermal, but because even an extreme temperature such as $-60°C = 213$ K is only about 20% different than the average temperature on the absolute scale, this isothermal model is pretty accurate: the observed pressure and density at $z = 10$ km are only about 20% lower than the model would predict.

ANSWERS TO EXERCISES

T5X.1 Helium atoms have a mass of $M_A = 4$ g per Avogadro's number. Therefore, $\varepsilon/k_B T$ in this case is

$$\frac{\varepsilon}{k_B T} = \frac{h^2}{8mL^2 k_B T}$$

$$= \frac{(6.63 \times 10^{-34} \text{ J·s})^2 (6.02 \times 10^{23})}{8(0.004 \text{ kg})(10^{-6} \text{ m})^2 (1.38 \times 10^{-23} \text{ J/K})(295 \text{ K})}$$

$$= 2.03 \times 10^{-9} \frac{\text{J·s}^2}{\text{kg·m}^2}\left(\frac{1 \text{ kg·m}^2/\text{s}^2}{1 \text{ J}}\right) \approx 2 \times 10^{-9} \tag{T5.45}$$

T5X.2 With $Z = (\pi/4\varepsilon)^{3/2}\beta^{-3/2}$, we have

$$Z = -\frac{1}{Z}\frac{dZ}{d\beta} = -\left(\frac{4\varepsilon\beta}{\pi}\right)^{3/2}\left(\frac{\pi}{4\varepsilon}\right)^{3/2}\frac{d}{d\beta}\beta^{-3/2}$$

$$= -\beta^{3/2}\left(-\frac{3}{2}\right)\beta^{-5/2} = \frac{3}{2\beta} = \frac{3}{2}k_B T \tag{T5.46}$$

T5X.3 According to the ideal gas law, the number of molecules is

$$N = \frac{PV}{k_B T} = \frac{(101,000 \text{ N/m}^2)(1400 \text{ cm}^3)}{(1.38 \times 10^{-23} \text{ J/K})(295 \text{ K})}\left(\frac{1 \text{ m}}{100 \text{ cm}}\right)^3$$

$$= 3.47 \times 10^{22} \frac{\text{N·m}}{\text{J}}\left(\frac{1 \text{ J}}{1 \text{ N·m}}\right) = 3.47 \times 10^{22} \tag{T5.47}$$

Since the mass of Avogadro's number of nitrogen molecules is $M_A = 28$ g/mol, the mass of our sample, according to equation T5.31, is

$$M = \frac{M_A}{N_A}N = \frac{28 \text{ g/mol}}{6.02 \times 10^{23}/\text{mol}}3.47 \times 10^{22}$$

$$= 1.62 \text{ g} \tag{T5.48}$$

T6 Distributions

Chapter Overview

Section T6.1: Counting Quantum States

This section argues that if we treat gas molecules as independent quantum particles in a cubical box of volume V (whose sides have length $L = V^{1/3}$), then their momentum components are quantized in units of $h/2L$, where m is the molecule's mass and h is Planck's constant. In a macroscopic box, $h/2L$ is so small that the momentum components are essentially continuous. The total number of states within a tiny range of width dp centered on a momentum magnitude $p = |\vec{p}|$ then turns out to be

$$n(p) = \frac{4\pi V}{h^3} p^2 \, dp \tag{T6.2}$$

Section T6.2: The Maxwell-Boltzmann Distribution

The probability that a given (nonrelativistic) gas molecule will be in a quantum state where it has speed $v = |\vec{v}|$ is proportional to the Boltzmann factor $e^{-mv^2/2k_B T}$. The result from the previous section tells us that the number of states that have the same speed v within $\pm\frac{1}{2}dv$ is proportional to $v^2 dv$, meaning that the probability that a given molecule has speed v within this range is proportional to $v^2 \, dv \, e^{-mv^2/2k_B T}$. We can determine the constant of proportionality by requiring that the probability be 1 that the molecule has a speed between 0 and ∞. The result is that

$$\text{Pr(speed within } dv \text{ centered on } v) = \mathcal{D}(v)\frac{dv}{v_P} \tag{T6.7a}$$

$$\text{where} \quad \mathcal{D}(v) = \frac{4}{\pi^{1/2}}\left(\frac{v}{v_P}\right)^2 e^{-(v/v_P)^2} \quad \text{and} \quad v_P \equiv \left(\frac{2k_B T}{m}\right)^{1/2} \tag{T6.7b}$$

- **Purpose:** This equation describes the probability that a molecule in a gas with temperature T has a speed within a range dv centered on speed v, where m is the molecule's mass, k_B is Boltzmann's constant, and v_P is the constant with units of speed defined in equation T6.7b.
- **Limitations:** The range dv must be small compared to v_P, since $\mathcal{D}(v)$ varies significantly over any range that is a significant fraction of v_P.
- **Notes:** The function $\mathcal{D}(v)$ is a version of the **Maxwell-Boltzmann distribution function** for molecular speeds in a gas. In general, distribution functions like $\mathcal{D}(v)$ are physically meaningful only when *integrated* over some (tiny or large) range. Note also that v_P does not depend on v or dv, but it does depend on temperature T and the molecular mass m.

$\mathcal{D}(v)$ is an example of a **distribution function,** whose goal in life is to be integrated. We can "integrate it" over an infinitesimal range simply by multiplying it by dv/v_P. To find the probability that the molecule's speed is within a finite range, we multiply $\mathcal{D}(v)$ by dv/v_P and actually integrate the product over that range. The web app MBoltz (at physapps.pomona.edu) can do this numerically.

Section T6.3: The Photon Gas

We can apply the same kind of reasoning to find the distribution of photon energies in a photon gas in a box whose walls have absolute temperature T. But there are two crucial differences between a photon gas and an ordinary gas. One is that a photon with momentum magnitude p has energy pc instead of $p^2/2m$. More consequentially, the number of photons in a box depends on the temperature T. This means that we cannot choose our quantum system to be a single photon, because that photon may appear or disappear at any time.

The trick is to choose our "quantum system" to be a quantum state, which may be occupied by an integer number of photons. The quantum state's energy is therefore $E = npc$ where p is the quantized momentum magnitude associated with the state and n is the number of photons that share that state. We can then use the result from section T6.1 to determine the distribution of photon energies. We find that the photon gas's total thermal energy density (energy per volume) U/V is proportional to T^4.

Section T6.4: Blackbody Emission

All objects with a nonzero absolute temperature T in fact radiate electromagnetic energy in the form of photons. We can link the characteristics of this radiated energy with those of the photon gas as follows. We first consider the energy that emerges from a small hole of area A poked in a box containing a photon gas. These emitted photons will have the same spectral distribution as the photon gas, and will have a power (energy per time) $P = A(U/V)(c/4)$. We next see that a perfectly absorbing surface (a **blackbody**) with the same temperature T facing such a hole must emit power per unit area at exactly the same rate as the hole for the two to be in thermal equilibrium (as they must be if they have the same temperature). The spectral distribution of photon energies must also be the same. An object that is not perfectly absorbing, but rather absorbs only a fraction ϵ of the light falling on it, must also emit the same fraction ϵ of the power that a blackbody would. If this **emissivity** ϵ varies with photon energy, then the power emitted at *each photon energy* is ϵ for that photon energy times the power that a blackbody would emit at that energy.

The power emitted by any object at absolute temperature T thus is given by

$$P = \epsilon \sigma A T^4 \quad \text{where} \quad \sigma = \frac{2\pi^5 k_B^4}{15h^3 c^2} = 5.67 \times 10^{-8} \frac{\text{W}}{\text{m}^2\text{K}^4} \tag{T6.25}$$

$$\frac{dP(\varepsilon)}{P_b} = \epsilon(\varepsilon)\,\mathcal{D}_{\text{ph}}(\varepsilon)\frac{d\varepsilon}{k_B T} \quad \text{where} \quad \mathcal{D}_{\text{ph}}(\varepsilon) \equiv \frac{15}{\pi^4}\frac{(\varepsilon/k_B T)^3}{e^{\varepsilon/k_B T} - 1} \tag{T6.26}$$

$$\varepsilon_P = 2.82 k_B T \tag{T6.27}$$

- **Purpose:** Equation T6.25 describes the power P (energy per time) that an object with absolute temperature T, surface area A, and emissivity ϵ emits in the form of electromagnetic wave energy (photons), where k_B is Boltzmann's constant, h is Planck's constant, and c is the speed of light. Equation T6.26 expresses the fraction of the emitted power that is carried by photons with energies within a range $d\varepsilon$ centered on ε compared to the total blackbody power P_b at the same T. Equation T6.27 specifies the most probable photon energy ε_P.
- **Limitations:** These equations do not work well for photons trapped in sub-micron-sized boxes, but otherwise are quite exact.
- **Notes:** An object's emissivity ϵ ranges from 0 for a perfect reflector to 1 for a perfect blackbody. Try not to confuse ϵ with photon energy ε. Again, the distribution function $\mathcal{D}_{\text{ph}}(\varepsilon)$ is meaningful only when it is integrated over some (tiny or large) range.

These are important equations with many applications. You can use the web app Planck (also at physapps.pomona.edu) to integrate the **Planck spectral energy distribution function** $\mathcal{D}_{\text{ph}}(\varepsilon)$ numerically over any finite range you specify.

T6.1   Counting Quantum States

In the last chapter, I mentioned that molecules in a gas do not all have the same speeds, but rather have a certain *distribution* of speeds. We can use the Boltzmann factor to determine exactly what this distribution looks like, and the results are pretty interesting. But the first step toward defining this distribution is to note that molecules in different quantum states (corresponding to different directions of motion) may have the same energy.

The quantum states for a molecule in a cubical box

Consider again a gas confined to a cubical box whose sides have length L. In section T5.2, we saw that in such a case, the absolute values of each of the three components of a molecule's momentum are *separately* quantized:

$$|p_x| = \left(\frac{h}{2L}\right)n_x, \quad |p_y| = \left(\frac{h}{2L}\right)n_y, \quad |p_z| = \left(\frac{h}{2L}\right)n_z \qquad \text{(T6.1)}$$

where n_x, n_y, and n_z are positive and completely independent integers.

Plotting the states in "momentum space"

The molecule's possible momentum values are thus equally separated in each component direction. We can better visualize this if we suppose (for the moment) that the molecule can move in only two dimensions. Suppose that we plot the allowed momentum component magnitudes on a graph whose axes are $|p_x|$ and $|p_y|$, respectively. The allowed momentum values will correspond to an evenly spaced square lattice of dots, as shown in figure T6.1. (Note that there are no dots along the $|p_x|$ and $|p_y|$ axes, because states where either n_x, n_y, or n_z are zero are not possible.) Each quantum state sits at the center of a square whose "area" on this diagram is $(h/2L)^2$. I have put "area" in quotation marks because the plane of the figure is not in ordinary space but rather in what we might call "momentum space," and the "area" has units of momentum squared, not meters squared. But perhaps you can see the analogy.

For a molecule moving in *three* dimensions, we would have a *cubical* lattice of equally spaced dots, which we can easily *imagine* extending above the plane of figure T6.1, but which would be quite hard to draw. In this case, each allowed momentum state occupies a cube with "volume" $(h/2L)^3$.

Now, states corresponding to the same kinetic energy $K = |\vec{p}|^2/2m$ (where m is the molecule's mass) will have the same value of $|\vec{p}|^2 = p_x^2 + p_y^2 + p_z^2$. Neighboring states will rarely have *exactly* the same values of $p_x^2 + p_y^2 + p_z^2 = h^2(n_x^2 + n_y^2 + n_z^2)/4L^2$ because n_x, n_y, and n_z are integers, and their squares

Figure T6.1

The dots indicate possible momentum states for a molecule moving in two dimensions in a square box of length L. The corresponding dots for a molecule moving in three dimensions would lie in a cubical lattice that would extend above the diagram's plane. Note (as mentioned in the last chapter) that each state embraces *both* possible directions of particle motion along each axis direction, so the quantum state corresponding to a given value of $|p_x|$ (for example) is agnostic about whether the molecule is moving in the $+x$ or $-x$ direction.

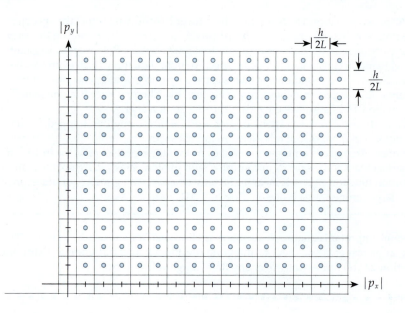

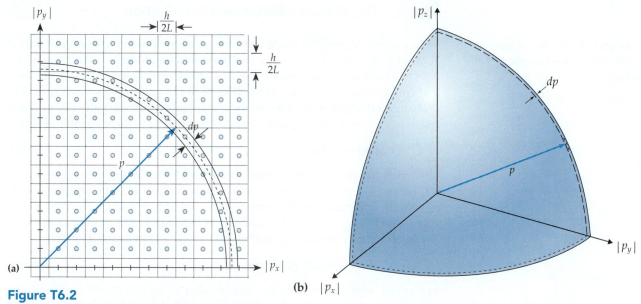

Figure T6.2

(a) For a molecule moving in two dimensions, the molecule's quantum states whose momentum magnitude is within the range $p \pm \frac{1}{2}dp$ lie within a quarter of a thin ring of radius p and thickness dp in the graph above. (b) For a molecule moving in *three* dimensions, such states lie within an eighth of a thin spherical shell of radius p and thickness dp on a similar three dimensional plot (consider the dots too tiny to see in this plot).

will only rarely add up to the same thing for neighboring states. But for a macroscopic container, $h/2L$ is an *extremely* tiny number (even compared to the typical momenta of individual gas molecules), so the dots are so finely spaced in practice that we can basically treat p_x, p_y, and p_z as if they were continuous variables. In such a case, instead of asking how many states have *exactly* the same value of $|\vec{p}|$, we might more practically ask how many states correspond to our molecule having a momentum magnitude lying between $|\vec{p}| \pm \frac{1}{2}d|\vec{p}|$, where $d|\vec{p}|$ is a very tiny range.

In this chapter (to reduce notational complexity), I am going to deviate from my usual practice and define $p \equiv |\vec{p}|$ and $v \equiv |\vec{v}|$. Please remember, though, that the simple letters p and v refer to *vector magnitudes*, and as such, always represent positive numbers.

For a molecule confined to move in two dimensions, allowed quantum states whose momentum magnitudes lie within the range $p \pm \frac{1}{2}dp$ will lie in one quarter of a thin ring on the $|p_x|/|p_y|$ plane whose central "radius" in our momentum space is p and whose width is dp (see figure T6.2a). Since each quantum state occupies the same "area" in this space, the number $n(p)$ of quantum states within the quarter-ring will be equal to the quarter-ring's "area" $\frac{1}{4}2\pi p\,dp$ divided by each state's "area" $(h/2L)^2$: $n(p) = (2\pi L^2/h^2)p\,dp$.

Now consider a molecule that can move in three dimensions. By analogy, its allowed quantum states corresponding to momentum magnitudes within the range $p \pm \frac{1}{2}dp$ will lie within an eighth of a thin spherical *shell* whose central "radius" is p and whose thickness is dp on a graph whose axes are $|p_x|$, $|p_y|$, and $|p_z|$ (see figure T6.2b). The number $n(p)$ of quantum states within this eighth of a shell will be its "volume" $\frac{1}{8}4\pi p^2\,dp$ divided by each state's "volume" $(h/2L)^3$:

$$n(p) = \frac{1}{8}\frac{8L^3}{h^3}4\pi p^2\,dp = \frac{4\pi V}{h^3}p^2\,dp \qquad (T6.2)$$

where $V = L^3$ is the cube's (and thus the gas's) true volume in space.

A notation convention for this chapter alone

Counting quantum states with momentum magnitudes in a certain tiny range.

T6.2   The Maxwell-Boltzmann Distribution

The probability that a gas molecule is in a quantum state where it has speed v

We are now in a position to answer this question: In a gas with a temperature T, what is the probability that a molecule will be moving at a certain speed $v \equiv |\vec{v}|$? Let's consider a single molecule to be our quantum system in thermal contact with a reservoir consisting of the remaining gas molecules. The probability that our molecule will be in a quantum state where it has (a nonrelativistic) speed $v = |\vec{p}|/m \equiv p/m$ is proportional to its Boltzmann factor:

$$\Pr(\text{state with speed } v) \propto e^{-E/k_BT} = e^{-p^2/2mk_BT} = e^{-mv^2/2k_BT} \tag{T6.3}$$

This gives the probability that the molecule will be in a single *quantum state* with speed v, but does not yet give us the probability that it simply *has* speed v, because the molecule has many quantum states that correspond to the same speed. To get the total probability of having speed v, we must multiply the Boltzmann factor by the number $n(v)$ of states having that speed:

$$\Pr(\text{molecule has speed } v) \propto n(v)e^{-mv^2/2k_BT} \tag{T6.4}$$

In the last section, we saw that the number of quantum states having the same momentum magnitude p within dp is $n(p) \propto p^2\,dp$. But since $v = p/m$, all states with the same p within dp will also have the same v within $dv = dp/m$. We can absorb the m into the constant of proportionality, so

$$\Pr(\text{within } dv \text{ around } v) \propto v^2\,dv\,e^{-mv^2/2k_BT} \tag{T6.5}$$

Note that $2k_BT/m$ has units of $(\text{J/K})(\text{K/kg}) = \text{m}^2/\text{s}^2$, so $v_P \equiv (2k_BT/m)^{1/2}$ is a constant with units of speed (I'll explain the subscript shortly). We can rewrite our expression for the probability using this constant as follows:

$$\Pr(\text{within } dv \text{ around } v) \propto v^2\,dv\,e^{-(v/v_P)^2} \propto \left(\frac{v}{v_P}\right)^2\left(\frac{dv}{v_P}\right)e^{-(v/v_P)^2} \tag{T6.6}$$

The beauty of the last version is that it is clearly a unitless number. Since the probability is also a unitless number, the constant of proportionality associated with the last expression must be unitless, too. We can determine that constant by requiring that the total probability that the molecule has *some* speed (*any* speed) be 1 (see problem T6D.1): the result turns out to be $4/\pi^{1/2}$. The complete expression for the probability is therefore

The Maxwell-Boltzmann distribution

$$\Pr(\text{speed within } dv \text{ centered on } v) = \mathcal{D}(v)\frac{dv}{v_P} \tag{T6.7a}$$

$$\text{where} \quad \mathcal{D}(v) = \frac{4}{\pi^{1/2}}\left(\frac{v}{v_P}\right)^2 e^{-(v/v_P)^2} \quad \text{and} \quad v_P \equiv \left(\frac{2k_BT}{m}\right)^{1/2} \tag{T6.7b}$$

- **Purpose:** This equation describes the probability that a molecule in a gas with temperature T has a speed within a range dv centered on speed v, where m is the molecule's mass, k_B is Boltzmann's constant, and v_P is the constant with units of speed defined in equation T6.7b.
- **Limitations:** The range dv must be small compared to v_P, since $\mathcal{D}(v)$ varies significantly over any range that is a significant fraction of v_P.
- **Notes:** The function $\mathcal{D}(v)$ is a version of the **Maxwell-Boltzmann distribution function** for molecular speeds in a gas. In general, distribution functions like $\mathcal{D}(v)$ are physically meaningful only when *integrated* over some (tiny or large) range. Note also that v_P does not depend on v or dv, but it does depend on temperature T and the molecular mass m.

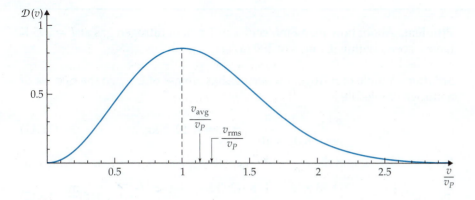

Figure T6.3

A graph of the Maxwell-Boltzmann distribution function $\mathcal{D}(v)$ as a function of v/v_P.

Experimental measurements of molecular velocities show that equation T6.7 does an excellent job of modeling those velocities.

Figure T6.3 shows a graph of the Maxwell-Boltzmann distribution function $\mathcal{D}(v)$. Note that at small values of v/v_P, the exponential is essentially equal to 1, so the function rises pretty much as $(v/v_P)^2$. For large values of v, the exponential takes over and reduces the probability back to zero. You can pretty easily show (see problem T6D.2) that the most probable speed, that is, the speed where $\mathcal{D}(v)$ is largest, is simply v_P (hence the subscript).

What the distribution looks like

Exercise T6X.1

Show that $v_P = 391$ m/s for oxygen (O_2) at 295 K. (O_2 has a mass of 32 g/mol.)

We can find the *average* speed v_{avg} by multiplying each possible speed by its probability and summing over all possible speeds. In the limit that the spacing between possible speeds is very small, the sum becomes an integral:

Different measures of a molecule's characteristic speed

$$v_{avg} = \int_0^\infty v\left[\mathcal{D}(v)\frac{dv}{v_P}\right] \tag{T6.8}$$

One can evaluate this by looking up the integral (see problem T6D.3): doing this yields

$$v_{avg} = \sqrt{\frac{8k_BT}{\pi m}} = \frac{2}{\sqrt{\pi}}\sqrt{\frac{2k_BT}{m}} = 1.13v_P \tag{T6.9}$$

This is larger than v_P because the function $\mathcal{D}(v)$ is not symmetric about its peak at v_P but has a long tail trailing off to $v = \infty$. This means that the average *should* be a bit higher than v_P. For oxygen at 295 K, $v_{avg} = 1.13(391 \text{ m/s}) = 442$ m/s. (Note that this is the average *speed*, not the average *velocity* $[\vec{v}]_{avg}$: the latter must be zero for a gas at rest!)

The so-called **root-mean-square** speed $v_{rms} \equiv ([v^2]_{avg})^{1/2}$ more heavily weights large speeds than v_{avg} does, so it is higher still. Because $\frac{1}{2}m[v^2]_{avg} = K_{avg} = \frac{3}{2}k_BT$, we have $[v^2]_{avg} = (2/m)\frac{3}{2}k_BT = 3k_BT/m$, implying that

$$v_{rms} \equiv \sqrt{[v^2]_{avg}} = \sqrt{\frac{3k_BT}{m}} = \sqrt{\frac{3}{2}}\sqrt{\frac{2k_BT}{m}} = 1.22v_P \tag{T6.10}$$

For oxygen at 295 K, $v_{rms} = 1.22(391 \text{ m/s}) = 477$ m/s.

These three quantities represent different ways of expressing the characteristic speed of molecules in a gas. While they are not identical, v_P and v_{rms} differ from the average speed by only a bit more than 10%.

Example T6.1

Problem: About how many molecules in 1 mol of nitrogen gas at $T = 295$ K have a speed within ± 1 m/s of 400 m/s?

Solution A mole of nitrogen molecules has a mass of 28 g, so the mass m of a nitrogen molecule is

$$m = \frac{0.028 \text{ kg}}{1 \text{ mol}} \left(\frac{1 \text{ mol}}{6.02 \times 10^{23}} \right) = 4.65 \times 10^{-26} \text{ kg} \tag{T6.11}$$

Let $v = 400$ m/s, and the range of acceptable speeds is $dv = 2$ m/s. Note that

$$v_P = \sqrt{\frac{2k_B T}{m}} = \sqrt{\frac{2(1.38 \times 10^{-23} \text{ J/K})(295 \text{ K})}{4.65 \times 10^{-26} \text{ kg}} \left(\frac{1 \text{ kg} \cdot \text{m}^2/\text{s}^2}{1 \text{ J}} \right)} = 418 \frac{\text{m}}{\text{s}} \tag{T6.12}$$

The range dv in this case is pretty small compared to v_P, so the integral over this tiny range will be approximately just $\mathcal{D}(v)(dv/v_P)$. We have

$$\left(\frac{v}{v_P} \right)^2 = \left(\frac{400 \text{ m/s}}{418 \text{ m/s}} \right)^2 = 0.916 \quad \text{and} \quad \frac{dv}{v_P} = \frac{2 \text{ m/s}}{418 \text{ m/s}} = 4.78 \times 10^{-3} \tag{T6.13}$$

$$\Rightarrow \quad \text{Pr(within } dv \text{ around } v) = \frac{4}{\pi^{1/2}} \left(\frac{v}{v_P} \right)^2 e^{-(v/v_P)^2} \frac{dv}{v_P}$$

$$= \frac{4}{\pi^{1/2}} (0.916) e^{-0.916} (4.78 \times 10^{-3}) = 3.95 \times 10^{-3} \tag{T6.14}$$

As we have 6.02×10^{23} molecules in our gas sample, we would expect about

$$(3.95 \times 10^{-3})(6.02 \times 10^{23}) = 2.38 \times 10^{21} \tag{T6.15}$$

molecules to have speeds in this range. Note that the probability comes out unitless (as it must). This also seems like a pretty reasonable fraction: *most of* the molecules will have speeds between roughly 0 and 800 m/s, and 2 m/s represents 2.5×10^{-3} of this range, so we might expect the probability to be about that order of magnitude.

Example T6.2

Problem: At $T = 295$ K, about what fraction of oxygen molecules in a sample of air will have a speed less than 200 m/s?

Solution Let $v_{\text{max}} = 200$ m/s. We know from exercise T6X.1 that $v_P = 391$ m/s for oxygen at $T = 295$ K, so the range of interest here is *not* small compared to v_P. In such a case, we must integrate $\mathcal{D}(v)$ over the range in question:

$$\text{Pr}(v < v_{\text{max}}) = \int_0^{v_{\text{max}}} \mathcal{D}(v) \frac{dv}{v_P} = \frac{4}{\pi^{1/2}} \int_0^{v_{\text{max}}} \left(\frac{v}{v_P} \right)^2 e^{-(v/v_P)^2} \frac{dv}{v_P}$$

$$= \frac{4}{\pi^{1/2}} \int_0^{x_{\text{max}}} x^2 e^{-x^2} dx \tag{T6.16a}$$

$$\text{where} \quad x \equiv \frac{v}{v_P} \quad \text{and} \quad x_{\text{max}} = \frac{v_{\text{max}}}{v_P} = \frac{200 \text{ m/s}}{391 \text{ m/s}} = 0.51 \tag{T6.16b}$$

The integral over x, alas, cannot be expressed in terms of simple functions, but we can evaluate it numerically (by dividing the x axis into very small steps, multiplying the step width dx by the function height $x^2 e^{-x^2}$ for each step, and then summing to find the area under the curve). I have written a web application called MBoltz that does exactly this: you can find it at http://physapps.pomona.edu/.

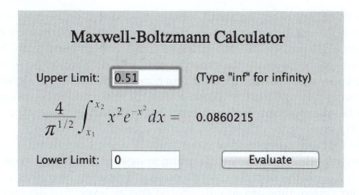

The figure above shows a screen shot of the application window when the program is correctly set up for this problem: I got the result displayed to the right of the integral by pressing the Evaluate button. We see that the probability that an oxygen molecule has a speed in this range is about 0.086, so about 8.6% of oxygen molecules have a speed smaller than 200 m/s at this temperature. This makes some sense: figure T6.3 shows that $\mathcal{D}(v)$ is not very large throughout most of the range in question.

T6.3 The Photon Gas

We saw in unit C that all objects with nonzero absolute temperature T emit energy in the form of electromagnetic waves. Just as a gas has a certain distribution of molecular energies, the photons (the quantum particles that carry the electromagnetic energy) emitted by an object have a certain characteristic distribution of energies at a given temperature T. We can determine this distribution much as we did for a gas in the last section.

To make the analysis a bit simpler, consider a collection of photons (a "photon gas") inside an otherwise empty cubical box whose sides have length L, and assume that these photons are in equilibrium with the box's walls at absolute temperature T. Photons are quantum particles just like molecules, so photons obey exactly the same mathematics described in section T6.1, meaning that the number of photon quantum states that share the same value of momentum magnitude p within a range dp is proportional to $p^2\,dp$.

However, a photon gas is different from an ordinary gas in two important ways. One is that photons are completely relativistic particles, while gas molecules are *not* relativistic. Therefore, a photon's energy E as a function of its momentum magnitude p is not given by the nonrelativistic expression $E = p^2/2m$, but rather the extreme relativistic result $E = pc$. Unit R proves this result, but if you haven't read unit R, let's simply accept this as true.

The other crucial difference is that *the number of photons in the box is not fixed*. Unlike gas molecules, which are not created or destroyed but rather simply bounce around inside the box, photons can be freely emitted into the photon gas or absorbed from the gas by the atoms in the container's walls. Indeed, we will see that the average number of photons in the box varies with the temperature T. This makes it impossible for us to take a single photon as our quantum system in equilibrium with the others, because a given photon (and thus our quantum system) may appear or disappear at any time!

However, the beauty of the mathematics developed in chapter T5 is that we can define our quantum system to be *anything we like*, as long as (1) it has well-defined energy states, (2) the multiplicity of those energy states is always 1,

Differences between a photon gas and an ordinary gas

and (3) its energy is always small compared with that of the reservoir with which it is in equilibrium. While choosing our quantum system to be a material object like a molecule is conceptually simple, we don't *have* to do this.

Choosing our "system" to be a specific quantum state instead of a molecule

So let us choose our quantum system in this situation to be something that we know *always* exists: the quantum *state* corresponding to a specific set of values of $|p_x|, |p_y|,$ and $|p_z|$. Now, you might find it odd to take a quantum *state* to be our quantum *system*, but such a system does satisfy the criteria above. (1) The energy of *one* photon with momentum $\vec{p}$ is $\varepsilon \equiv pc$. But it is possible that *any* number of photons share the same momentum state. So the energy levels of our chosen quantum system are simply $E_n = n\varepsilon$, where $n = 0, 1, 2, \ldots$ is the number of photons that happen to be in this quantum state. Our system thus has equally spaced energy levels that depend on the number of photons in that state. (2) All photons are completely identical, meaning that we can't tell which of the photons in the state is which. This means that rearrangements of the photons don't count as separate states: not only do all rearrangements *look* the same, they really *are* the same according to quantum physics. So the multiplicity of an energy level of our system remains 1 no matter how many photons are in the state. (3) Our photon gas actually involves an infinite number of possible quantum states, so any given state will always hold only a tiny energy relative to the whole gas.

Calculating the state's average energy

The partition function for our quantum system is the sum of the Boltzmann factors for all its energy states $E_n = n(pc)$:

$$Z = \sum_{n=0}^{\infty} e^{-pc\beta n} = \sum_{n=0}^{\infty} x^n = \frac{1}{1-x} = \frac{1}{1 - e^{-pc\beta}} \tag{T6.17}$$

where $\beta = 1/k_B T$ and $x \equiv e^{-pc\beta}$ and where I have used the identity first mentioned in equation T4.38 to evaluate the sum. Our system's average energy is

$$E_{avg} = -\frac{1}{Z}\frac{dZ}{d\beta} = \frac{pc}{e^{pc\beta} - 1} \tag{T6.18}$$

Exercise T6X.2

Verify that equation T6.18 is correct.

Calculating the photon gas's total energy

Now, in an ordinary gas, our quantum systems were identical molecules, so to get the gas's total energy U, we only needed to multiply the energy per molecule by the number of molecules N. In this case, our systems are quantum states, and they are *not* identical. So to find the gas's total energy, we must actually sum over all the different quantum states. Let $E(p)$ be the average energy for a given state with momentum magnitude p. We know from section T6.1 that the number of states having the same magnitude of p within a range of width dp is $n(p) = (4\pi V/h^3)p^2\,dp$ (see equation T6.2). Technically, we actually have twice this number of states, because photons have two independent polarization states (you can think of these as possible spin orientations). Therefore, the total thermal energy of our photon gas is

$$U = \sum_{all\ p} n(p)E(p) \approx \int_0^{\infty} \left(\frac{8\pi V}{h^3} p^2 dp\right)\left(\frac{pc}{e^{pc\beta} - 1}\right) \tag{T6.19}$$

We can put this in a somewhat more evocative form if we change variables to $u \equiv pc\beta = pc/k_B T$. Note that u is a unitless value that specifies the ratio of a photon's energy $\varepsilon = pc$ to $k_B T$, the characteristic energy associated with temperature T. Substituting $p = uk_B T/c$ and $dp = k_B T\,du/c$ into the integral and dividing both sides by the gas volume V yields

$$\frac{U}{V} = \frac{8\pi(k_B T)^4}{(hc)^3} \int_0^\infty \frac{u^3 \, du}{e^u - 1} \qquad (T6.20a)$$

If you look up the integral, you will find that its value is $\pi^4/15$. Therefore,

$$\frac{U}{V} = \frac{8\pi^5(k_B T)^4}{15(hc)^3} \qquad (T6.20b)$$

This expression gives the *energy density* of the photon gas (energy per unit volume instead of energy per molecule). Note (shockingly) that this energy density depends not *linearly* on $k_B T$ but rather varies as $(k_B T)^4$!

Now, note that the quantity $u = \varepsilon/k_B T$ is a unitless measure of photon energy (the photon energy ε as a fraction or multiple of $k_B T$): let's call u the **rescaled energy**. The fraction of the photon gas's energy carried by photons with rescaled energy u within du is

The Planck spectral distribution of photon energies

$$\frac{dU(u)}{U} = \mathcal{D}_{\rm ph}(u)\,du, \qquad \text{where} \qquad \mathcal{D}_{\rm ph}(u) = \frac{15}{\pi^4}\frac{u^3}{e^u - 1} \qquad (T6.21)$$

Just as the Maxwell-Boltzmann distribution $\mathcal{D}(v)$ for ordinary gases tells what fraction of molecules have speeds within a range dv centered on a given speed v, the **Planck spectral distribution** $\mathcal{D}_{\rm ph}(u)$ tells us what fraction of the photon gas's energy is carried by photons with rescaled energy within a range du centered on a given value u.

Figure T6.4 shows a plot of $\mathcal{D}_{\rm ph}(u)$ [note that it even *looks* like $\mathcal{D}(v)$!]. This function has a peak at $u \approx 2.82$: the most probable photon energy is thus

$$\varepsilon_P \approx 2.82\,k_B T \qquad (T6.22)$$

Wien's law

As one might expect from previous results, the most probable photon energy ε_P scales *linearly* with T: physicists call this **Wien's law**. The reason that the photon gas's total energy scales as T^4 is because the number of photons in the box scales as T^3 (see problem T6D.5)!

The web application Planck does the same kind of numerical integrations of the Planck spectral distribution that MBoltz does of the Maxwell-Boltzmann distribution: you can use it to find what fraction of the emitted energy is emitted within any finite range of rescaled energies u. You can also find this app at http://physapps.pomona.edu/.

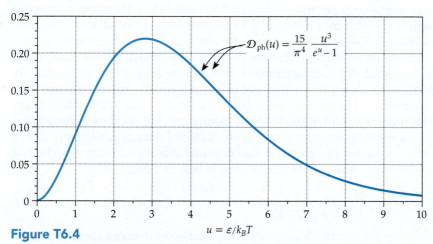

Figure T6.4

The Planck spectral distribution function $\mathcal{D}_{\rm ph}$ as a function of $u \equiv \varepsilon/k_B T$.

Figure T6.5
A photon gas in a box with a small hole of area A in it. If we were to pretend that all photons in the box happen to be moving to the right, then any photons emerging from the hole during a short time interval of Δt must have been somewhere within the dashed region at the beginning of the interval.

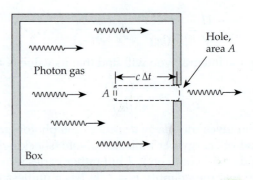

Calculating the rate at which electromagnetic energy escapes from the hole

T6.4   Blackbody Emission

We now know the characteristics of the photon gas *inside* a box. However, this also determines the characteristics of the photons emitted by an object at a certain absolute temperature T. In this section, we will see why.

First consider a cubical box with a small hole in it, as shown in figure T6.5. If the hole is sufficiently small, then its presence will not significantly disturb the equilibrium of the photon gas inside the box. Therefore, the photons that happen to emerge from the hole will have exactly the same spectral distribution as the photons in the box.

At what rate do photons carry energy out through the hole? Suppose for the moment that all photons in the box happen to be moving directly to the right. (This is totally absurd, but bear with me.) If this is so, all photons that escape through the hole during a time interval Δt must have (at the start of the interval) been inside the dashed region in figure T6.5 whose area is the hole's area A and whose length to the left is $c\,\Delta t$. The volume of this region is $Ac\,\Delta t$, so the energy that this region contains is $(U/V)Ac\,\Delta t$, implying that the power (energy per time) P that escapes through the hole must be

$$P \sim \frac{U}{V}\frac{Ac\,\cancel{\Delta t}}{\cancel{\Delta t}} = \frac{U}{V}Ac \tag{T6.23}$$

We might have arrived at this result simply by analyzing units. Doubling the photon gas's energy density U/V would plausibly double the energy emerging from the hole, as would doubling the hole area A, so the power must be proportional to $(U/V)A$. But $(U/V)A$ has SI units of $(J/m^3)m^2 = J/m$, while the power (energy per time) P emitted has units of J/s. Therefore, we need to multiply $(U/V)A$ by a constant with units of m/s, or speed, to get the units to work out. The only relevant speed in this problem is the photons' speed c.

Now, our assumption that the photons are all moving in the same direction is absurd. But both approaches above make it clear that accounting for arbitrary directions of motion will only multiply the result by some unitless constant. That constant must in fact be smaller than $1/2$, because at any given time, only half of the box's randomly moving photons in the box are actually moving toward the hole at all. A careful but not particularly illuminating calculation (see problem T6D.6) shows that the constant must be $1/4$. The power per unit area emitted by the hole is therefore

$$\frac{P}{A} = \frac{U}{V}\frac{c}{4} = \frac{c}{4}\left(\frac{8\pi}{h^3 c^3}\right)\frac{\pi^4}{15}(k_B T)^4 = \frac{2\pi^5 k_B^4}{15 h^3 c^2}T^4 = \sigma T^4 \tag{T6.24}$$

where $\sigma = 5.67 \times 10^{-8}\ \text{W}/(\text{m}^2\text{K}^4)$ is the **Stefan-Boltzmann** constant (which you can remember as "5-6-7-8": just don't forget the minus sign).

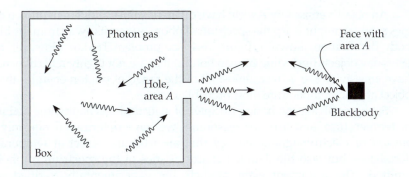

Figure T6.6
A perfectly absorbing blackbody with a face of area A facing a hole of area A in a cubical box containing a photon gas at temperature T. If the blackbody has the same temperature T, then it must emit exactly the same amount of energy as the hole to remain in equilibrium.

Now, what do photons emerging from a small hole in a cubical box have to do with photons emitted by a general object at temperature T? As we will see, basic thermal physics *requires* that these phenomena be connected.

Consider first an object that perfectly absorbs photons: we call such an object a **blackbody.** Suppose that such an object has a face of area A, which we place facing a hole with area A in a cubical box containing a photon gas (see figure T6.6). Because the face and hole have the same area, each must absorb an equal fraction of the other's emitted photons. Suppose also that the object and the photon gas in the box have the same temperature T.

Consider now the following argument by contradiction. Suppose that the blackbody's face emits *less* electromagnetic energy than the hole does. Then the blackbody would receive (and absorb, since it is perfectly black) more energy from the hole than it itself emits, meaning that its thermal energy would increase. But this is impossible! If the blackbody and the photon gas in the box have the same temperature T, then they must be in equilibrium *by definition*. For similar reasons, the blackbody cannot emit *more* energy than the hole does. So the definition of temperature requires that the blackbody emit *exactly* as much energy per unit area per unit time as the hole does!

This argument does not just apply to the *total* energy emitted by each object: it also refers to the exact *distribution* $\mathcal{D}_{\mathrm{ph}}(u)$ of photon energies described in figure T6.4. Why? Suppose we place an optical filter between the hole and blackbody that permits photons having only a certain energy to pass through. The blackbody and the photon gas emerging from the hole must *still* be in equilibrium, so the blackbody must emit exactly as much energy *at that particular wavelength* as the hole does. Because we could choose our filter to pass any photon energy we choose, the amount of electromagnetic energy emitted by the blackbody must be exactly like that emitted by the hole for every possible photon energy.

However, almost no object is perfectly absorbing, so now consider an object that scatters away (say) half of the photons that fall on it. The object must still be in thermal equilibrium with the hole, so because it absorbs only half the photons it receives from the hole, it must also *emit* only half the number of photons that the hole does. Physicists define an object's **emissivity** ϵ to be the fraction of electromagnetic energy falling on it that an object absorbs: this quantity ranges from 1 for a perfect blackbody to 0 for a perfect reflector ($\epsilon = 1/2$ in our example). We see that the rate at which a general object at absolute temperature T *emits* electromagnetic energy must therefore be

$$P = \epsilon \sigma A T^4$$

This is the **Stefan-Boltzmann law.** Even though we have derived it through nearly pure logic, this law is extremely successful experimentally. Indeed, it is one of the most useful laws of physics.

Emission by a perfectly absorbing "blackbody"

Emission by objects that are not blackbodies

The Stefan-Boltzmann law

Emissivity can be different at different photon energies

An object's emissivity ϵ *might* have different values at different photon energies ε, which might warp the spectrum of photons it emits away from the blackbody distribution shown in figure T6.4 (see problem T6M.6). However, most everyday objects (including human bodies) that are not highly reflective metals have emissivity $\epsilon \approx 1$ over almost all of the range of photon energies that an object at room temperature most probably radiates.

So, by applying the basic principles of statistical mechanics, the definition of temperature, and a bit of reasoning, we have determined not only the *amount* of electromagnetic energy that an arbitrary object at temperature T radiates, but also the detailed characteristics of the spectrum of photons emitted. The argument even predicts the experimentally verified *value* $2\pi^5 k_B^4/(15h^3c^2)$ of the Stefan-Boltzmann constant σ! To summarize:

The Stefan-Boltzmann law

The Planck spectral distribution for photon energies

Wien's law

$$P = \epsilon\sigma A T^4 \quad \text{where} \quad \sigma = \frac{2\pi^5 k_B^4}{15h^3c^2} = 5.67 \times 10^{-8} \, \frac{\text{W}}{\text{m}^2\text{K}^4} \tag{T6.25}$$

$$\frac{dP(\varepsilon)}{P_b} = \epsilon(\varepsilon)\mathcal{D}_{\text{ph}}(\varepsilon)\frac{d\varepsilon}{k_B T} \quad \text{where} \quad \mathcal{D}_{\text{ph}}(\varepsilon) \equiv \frac{15}{\pi^4}\frac{(\varepsilon/k_B T)^3}{e^{\varepsilon/k_B T} - 1} \tag{T6.26}$$

$$\varepsilon_P = 2.82\, k_B T \tag{T6.27}$$

- **Purpose:** Equation T6.25 describes the power P (energy per time) that an object with absolute temperature T, surface area A, and emissivity ϵ emits in the form of electromagnetic wave energy (photons), where k_B is Boltzmann's constant. Equation T6.26 expresses the fraction of the emitted power that is carried by photons with energies within a range $d\varepsilon$ centered on ε compared to the total blackbody power P_b at the same T. Equation T6.27 specifies the most probable photon energy ε_P.
- **Limitations:** These formulas do not work well for photons trapped in sub-micron-sized boxes, but otherwise are quite exact.
- **Notes:** An object's emissivity ϵ ranges from 0 for a perfect reflector to 1 for a perfect blackbody. Try not to confuse ϵ with photon energy ε. Again the distribution function $\mathcal{D}_{\text{ph}}(\varepsilon)$ is meaningful only when it is integrated over some (tiny or large) range.

(Equation T6.26 follows from equation T6.21 because $P = \frac{1}{4}(U/V)Ac$ (for energy emitted by a hole of area A in a cubical box of volume V) implies that $dP/P = dU/U$, and I have simply substituted $\varepsilon/k_B T$ and $d\varepsilon/k_B T$ for u and du in the earlier equation.)

These are very powerful principles whose applications range from the power radiated by nanoscale devices to the entirety of the early universe. The homework problems in this chapter outline just a few such applications.

Exercise T6X.3

Consider an object whose absolute temperature is $T = 1500$ K (5 times room temperature). **(a)** What is the most probable photon energy? **(b)** Use the Planck web app to determine the fraction of the energy emitted by this object that is in the visible range ($\varepsilon = 1.77$ eV to 3.1 eV). (*Hint:* $k_B T_{\text{room}} = 0.0254$ eV.)

TWO-MINUTE PROBLEMS

T6T.1 When *counting* particle-in-a-box quantum states in three dimensions, why do we divide the "volume" of only one eighth of a full spherical shell by the "volume" per quantum state? Why not the full shell's volume?
A. Molecules in a gas can only have positive momentum components.
B. A single quantum *energy* state embraces both positive and negative values for each momentum component.
C. A particle's momentum magnitude must be positive.
D. The factor of 2 in the expression $h/2L$ already counts both possible momentum signs.
E. We really do include the full shell: the diagram only shows 1/8 of the shell to make it easier to draw.
F. Some other reason (specify).

T6T.2 *Physically*, why is a gas molecule's speed v more likely to be near v_P than near zero?
A. There are fewer quantum states with smaller speeds than larger speeds.
B. The Boltzmann factor has a larger value near $v = v_P$ than near $v = 0$.
C. We know that $\frac{1}{2}m[v^2]_{avg} = \frac{3}{2}k_B T$, and v_P is closer to $[v^2_{avg}]^{1/2}$ than zero is.

T6T.3 *Physically*, why is a gas molecule's speed v more likely to be near v_P than near $2v_P$?
A. There are fewer quantum states for larger speeds than there are at speeds near v_P.
B. The Boltzmann factor is larger at v_P than at $2v_P$.
C. We know that $\frac{1}{2}m[v^2]_{avg} = \frac{3}{2}k_B T$, and v_P is closer to $[v^2_{avg}]^{1/2}$ than $2v_P$ is.

T6T.4 How should $([v^3]_{avg})^{1/3}$ compare to $([v^2]_{avg})^{1/2}$ for an ideal gas?
A. $([v^3]_{avg})^{1/3} > ([v^2]_{avg})^{1/2}$
B. $([v^3]_{avg})^{1/3} = ([v^2]_{avg})^{1/2}$
C. $([v^3]_{avg})^{1/3} < ([v^2]_{avg})^{1/2}$
D. One cannot tell without careful calculation.

T6T.5 Which of the speeds v listed below is the *lowest* speed where the probability of a gas molecule having that speed is negligible compared to that for $v = v_P$ ("negligible" meaning less than 1/100 of its value at $v = v_P$).
A. $v \approx 0.2v_P$
B. $v = 2v_P$
C. $v = 3v_P$
D. $v = 5v_P$

T6T.6 How would the values of v_P for carbon dioxide gas compare to that of nitrogen gas at a given temperature? (*Hint:* Carbon dioxide gas has a molar mass of 44 g/mole; nitrogen has a molar mass of 28 g/mole.)
A. v_P will be larger for carbon dioxide.
B. v_P will be larger for nitrogen.
C. v_P will be equal in both cases.
D. One does not have enough information to tell for sure.

T6T.7 Will a blackbody at temperature T emit more photons per second with energy $0.5k_B T$ or with energy $1.5k_B T$? (Note: The amount of energy emitted per time is not the same as the number of photons emitted per time, though these quantities are related.)
A. It will emit more photons with the lower energy.
B. It will emit more photons with the higher energy.
C. The number of photons emitted will be the same.
D. One has no way of telling.

T6T.8 Will a blackbody at temperature T emit more photons per second with energy $k_B T$ or with energy $6k_B T$? (Note: The amount of energy emitted per time is not the same as the number of photons emitted per time, though these quantities are related.)
A. It will emit more photons with the lower energy.
B. It will emit more photons with the higher energy.
C. The number of photons emitted will be the same.
D. One has no way of telling.

T6T.9 Doubling a blackbody's temperature doubles the number of photons it emits per second. T or F?

T6T.10 An object strongly reflects red light, but readily absorbs all other colors of light. Relative to the number of thermal photons of red light emitted by a black body at the same temperature T, our object's thermal emission rate of photons of *red* light will be
A. enhanced.
B. suppressed.
C. unchanged.
D. The answer depends on the value of T.

T6T.11 Suppose that a star has twice the surface temperature of the sun and also twice the radius. How many times larger is the rate at which this star emits energy compared to the rate at which the sun emits energy?
A. The star radiates at the same rate as the sun.
B. 2 times
C. 4 times
D. 8 times
E. 16 times
F. 32 times
T. 64 times

T6T.12 Suppose that we double the electrical power flowing into the filament of an incandescent bulb. By what factor will the most probable photon energy ε_P increase?
A. $2^{1/4}$
B. $2^{1/3}$
C. $2^{1/2}$
D. 2
E. 4
F. 8
T. 16

HOMEWORK PROBLEMS

Basic Skills

T6B.1   Consider nitrogen gas in a container at room temperature $T = 295$ K. A molecule's average kinetic energy is $K_{avg} = \frac{3}{2}k_B T$. Calculate the momentum magnitude p of a nitrogen molecule having this kinetic energy. (*Hints:* A mole of nitrogen has a mass of 28 g. Your answer should be on the order of 10^{-23} kg·m/s.)

T6B.2   Consider a gas in a cubical container 0.1 m on a side. How many states lie within $\pm 0.5\%$ of the momentum magnitude $p = 2.0 \times 10^{-23}$ kg·m/s (a plausible momentum for a nitrogen molecule in a gas at room temperature)?

T6B.3   Consider a gas in a cubical container 1.0 cm on a side. How many states lie within $\pm 0.01\%$ of the momentum magnitude $p = 1.6 \times 10^{-23}$ kg·m/s (a plausible momentum for a nitrogen molecule in a gas at room temperature)?

T6B.4   What is the probability that an oxygen molecule in a gas at room temperature has a speed in the range 500 m/s $\pm$ 5 m/s?

T6B.5   Consider nitrogen gas at 0°C.
(a) Show that $v_P = 403$ m/s.
(b) What is the probability that a nitrogen molecule in this gas has a speed in the range 250 m/s $\pm$ 2 m/s?

T6B.6   Use the MBoltz web app to find the probability that an oxygen molecule in a gas at room temperature (295 K) will have a speed between 300 and 500 m/s.

T6B.7   Use the MBoltz web app to find the probability that a gas molecule will have a speed greater than v_P.

T6B.8   Consider a mixture of helium gas and oxygen gas in the same container.
(a) Is v_P the same for both types of molecule? If not, which gas has the greater value for v_P?
(b) For which gas is the probability greater that a molecule has a speed greater than $3v_P$ (or is this probability the same for both gases)?

T6B.9   What is the energy density U/V of a photon gas at
(a) room temperature ($T = 295$ K), and
(b) at a temperature ten times larger?

T6B.10   The sun's surface temperature is about 5800 K.
(a) About how much electromagnetic wave energy does a cubic meter of space near the sun's surface contain?
(b) What is the most probable photon energy ε_P (in eV) for photons emitted by the sun?

T6B.11   If a blackbody has a temperature of 373 K, what fraction of the electromagnetic power that it emits consists of photons with energies in the range of 0.50 eV $\pm$ 0.01 eV?

T6B.12   A person has a surface temperature $\approx 37°C = 310$ K.
(a) What is the most probable photon energy ε_P (in eV) for photons emitted by a person?
(b) Photons of visible light have energies that range from 1.77 eV (deep red) to 3.1 eV (deep violet). How does the most probable photon energy that you calculated for part (a) compare to this range?
(c) Use the Planck web app to determine what fraction of the energy emitted by a human body is visible.

T6B.13   Use the Planck web app to determine what fraction of a photon gas's energy is carried by photons having an energy ε greater than $10k_B T$.

T6B.14   A red dwarf star might have a surface temperature of 2950 K (about 10 times room temperature). Use the Planck web app to determine what fraction of the energy emitted by such a star falls in the visual range (photon energies 1.77 eV to 3.1 eV).

T6B.15   The value of $\mathcal{D}_{ph}(\varepsilon)$ is about the same at $\varepsilon = k_B T$ as it is at $\varepsilon = 6k_B T$. Calculate the ratio in the number of photons emitted per second within $\pm d\varepsilon$ of these energies, and explain why this ratio is *not* one.

T6B.16   Calculate the number of thermal photons inside a cubical box that is 0.1 m on a side and is at room temperature. (*Hint:* See problem T6D.5.)

Modeling

T6M.1   Suppose that in a certain physics experiment, a plasma of ionized oxygen atoms emerges from an opening in a furnace with a temperature of 3200 K. We then use a velocity selector (such as the one discussed in unit E) that passes only those ions whose speeds are within ± 0.01 km/s of 1.40 km/s. If we need about 2×10^{12} ions per second with that speed for our experiment, about how many ions per second must the furnace emit? (We can use this information to help determine what furnace to order.)

T6M.2   Water molecules evaporate from the surface of a puddle when random collisions give them a kinetic energy that exceeds the binding energy holding the molecules together, which is about 0.42 eV. As a first approximation, assume that the water molecules have speeds given by the Maxwell-Boltzmann distribution. About what fraction of molecules on the surface can evaporate if the puddle has a temperature of 20°C? What is the fraction if the temperature is 50°C? Use this to qualitatively explain why when you rub your hands under a bathroom hand dryer, most of the drying seems to happen during the last few seconds.

T6M.3   *Without* using MBoltz, estimate the fraction of molecules in a gas at temperature T whose speed is less than 0.1 times the most probable speed at that temperature.

(*Hint:* It is *not* a good approximation to assume that this range of speeds is small enough that we can just multiply $\mathcal{D}(v)$ by $dv = 0.1v_P$. Why not? So you will have to do an integral over $\mathcal{D}(v)$. But you can replace the exponential by something simpler over this range. You can *check* your work with MBoltz.)

T6M.4   In this problem, we will consider the energy budget required for a human body to emit thermal photons.

(a) Consider a relatively small person with a surface area of 1.0 m^2 and a body temperature of $37°C = 310$ K. Show that such a person would need to consume about 11,000 food calories a day simply to replenish the energy emitted by his or her body. (Assume that a person's emissivity $\epsilon \approx 1$ in the relevant photon energy range.)

(b) Of course, this is for a person who is naked in a vacuum. A small person actually requires something more like 2000 food calories a day to maintain body temperature. This is partly because a person's typical surroundings emit thermal photons back at the person. Calculate the *net* energy per day (in food calories) that a naked person in a room at 295 K would emit. (*Hint:* Equation T6.25 also describes the power that a surface *absorbs* from surroundings at absolute temperature T.)

(c) Clothing also helps. Suppose that the person's entire surface area A is surrounded by a layer of clothing that completely absorbs the person's thermal photons, but also emits photons in both directions from its surface, half back to the person from the layer's inner surface and half to the surroundings from its outer surface. Let P_p/A be the power per unit area emitted by the person, let P_c/A be the power per unit area emitted by *each surface* of the clothing layer, and let P_s/A be the power per unit area that the layer absorbs from the room. When the clothing layer is in equilibrium with both the person at 310 K and its surroundings at 295 K, then the net energy flow to that layer must be zero, meaning that $P_p/A + P_s/A = 2P_c/A$ (since the layer has *two* surfaces of area A that are both emitting). Show that the layer must have absolute temperature $T_c = 302.8$ K.

(d) Calculate, therefore, the *net* rate at which a clothed person loses energy in food calories per day.

T6M.5   Use some kind of computer tool to construct plots of $(1/A)(dP(\varepsilon)/d\varepsilon)$ (the blackbody's radiated power per unit area per unit energy range) as a function of photon energy ε (in eV) for photon gases at 1000 K, 1500 K, and 2000 K. Draw all three plots on the same graph, and comment about how the three graphs compare.

T6M.6   Suppose that we paint a certain object with red paint so that it absorbs all light falling on it *except* for red light in a photon energy range between 1.8 eV and 2.0 eV, which it completely reflects. Suppose now that this object has a temperature such that $\varepsilon_P = 1.40$ eV. Sketch a graph of $dP(\varepsilon)/P_b$ for the energy emitted by this object (don't worry about the numbers on the vertical axis), and carefully explain why it *cannot* look the same as figure T6.4 if the object is to remain in thermal equilibrium in all cases.

T6M.7   Suppose I want to skew the color of the photon gas in a container, so I go inside the container and quickly paint its inside walls with green paint that *almost* completely reflects green light corresponding to photon energies between 2.4 eV and 2.5 eV. Will this change the energy distribution of photons in the photon gas? If so, will the gas become less green or more green, and why? If not, why not?

Derivation

T6D.1   The total probability of a molecule having a speed between zero and infinity must be 1. By looking up the integral online (or using WolframAlpha), evaluate the constant A in the expression

$$1 = \int_0^\infty \mathcal{D}(v)\frac{dv}{v_P} = \int_0^\infty A\left(\frac{v}{v_P}\right)^2 e^{-(v/v_P)^2}\left(\frac{dv}{v_P}\right) \quad \text{(T6.28)}$$

(*Hint:* Define $x = v/v_P$ before looking up the integral.)

T6D.2   Prove that the maximum of the Maxwell-Boltzmann distribution function $\mathcal{D}(v)$ is precisely at $v = v_P$.

T6D.3   The equation below (equation T6.8) gives the average value of a molecule's speed in a gas:

$$v_{\text{avg}} = \int_0^\infty v\left[\mathcal{D}(v)\frac{dv}{v_P}\right] = v_P\int_0^\infty \left(\frac{v}{v_P}\right)\mathcal{D}(v)\frac{v}{v_P} \quad \text{(T6.29)}$$

By looking up the integral online (or using WolframAlpha, show that $v_{\text{avg}} = \sqrt{(4/\pi)}\, v_P = \sqrt{8k_BT/\pi m}$ (equation T6.9). (*Hint:* Substitute in $x = v/v_P$ before looking up the integral.)

T6D.4   The equation below gives the average value of the square of a molecule's speed:

$$[v^2]_{\text{avg}} = \int_0^\infty v^2\left[\mathcal{D}(v)\frac{dv}{v_P}\right] = v_P^2\int_0^\infty \left(\frac{v}{v_P}\right)^2 \mathcal{D}(v)\frac{v}{v_P} \quad \text{(T6.30)}$$

(a) By looking up the integral online (or using Wolfram-Alpha), find the value of $[v^2]_{\text{avg}}$. (*Hint:* Substitute in $x = v/v_P$ before looking up the integral.)

(b) Explain why your answer makes sense. (*Hint:* What do we know about K_{avg}?)

T6D.5   In this problem, we will calculate the number of photons in a photon gas at absolute temperature T.

(a) The number of photons whose energies lie within the range $\pm\frac{1}{2}d\varepsilon$ about some energy ε is equal to the total photon energy in that range divided by ε, right? Use this to set up the integral that you need to calculate the number of photons per unit volume N/V in the gas.

(b) If you change variables to integrate over the unitless variable $u = \varepsilon/k_BT$, you should see that N/V depends on $(k_BT)^3$. Explain why knowing the specific value of the integral over u is irrelevant for this claim.

(c) By looking up the integral (perhaps on Wikipedia's list of integrals), show that the number of photons per unit volume in a photon gas with temperature T is

$$\frac{N}{V} = 8\pi \cdot 2.404\left(\frac{k_BT}{hc}\right)^3 = 60.42\left(\frac{k_BT}{hc}\right)^3 \quad \text{(T6.31)}$$

T6D.6 (Adapted from Schroeder, *Thermal Physics*, 2000.) We can show that the power P of electromagnetic energy emerging from a hole of area A in the wall of a container holding a photon gas of energy density U/V is given by $P = (c/4)(U/V)A$ as follows. Consider a certain time interval of length dt centered at $t = 0$ during which photons emerge from the hole. Because all photons move at the speed of light, at some earlier time $t = -T$, all the photons that emerged from the hole must have been within a spherical shell of radius $R = Tc$ and thickness $dR = c\,dt$ centered on the hole, as shown below:

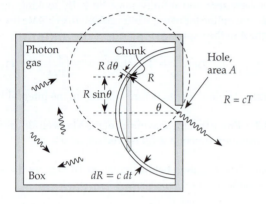

(a) Consider a small chunk of this shell, as shown in the diagram. Of all the photons that are in the chunk at time $t = -T$, the fraction that make it through the hole must be the same as the ratio of the hole's apparent area (as viewed from the chunk) to the entire area of a sphere of radius R centered on the chunk (the dashed circle on the diagram). But to an observer on the chunk, the hole's apparent area is not A but rather has been foreshortened to $A\cos\theta$ (note that the hole's apparent area is A when viewed from $\theta = 0$ but goes to zero as θ approaches 90°). Argue then that if the chunk's volume is dV, then the portion dE of the chunk's energy that makes it through the hole must be

$$dE = \frac{A\cos\theta}{4\pi R^2}\left(\frac{U}{V}\right)dV \qquad (T6.32)$$

(b) All of the chunks in the shaded ring are the same angle θ from the dashed line. This ring has cross-sectional area $(R\,d\theta)(c\,dt)$ and radius $R\sin\theta$. Argue then that the total rate at which energy flows through the hole due to all the chunks on this ring is simply

$$P_{ring} = \frac{dE_{ring}}{dt} = \frac{A\cos\theta}{2}\left(\frac{U}{V}\right)c\sin\theta\,d\theta \qquad (T6.33)$$

(c) To find the total power emitted by all rings, we must sum this over all rings. In the limit that $d\theta$ goes to zero, this sum becomes the integral

$$P = \frac{A}{2}\left(\frac{U}{V}\right)c\int_0^{\pi/2}\sin\theta\cos\theta\,d\theta \qquad (T6.34)$$

This is a fairly simple integral. Integrate it to show that $P = (c/4)(U/V)A$, as claimed in the text. (*Hint:* Change the integration variable from θ to $u = \sin\theta$.)

T6D.7 Prove that the peak of the Planck distribution $\mathcal{D}(\varepsilon)$ occurs at $\varepsilon/k_B T = 2.82$. (Solve the transcendental equation you get by using WolframAlpha or trial and error.)

T6D.8 The energy in electromagnetic waves with wavelength λ is carried by photons with energy $\varepsilon = hc/\lambda$. Let's define a "rescaled wavelength" $w \equiv \lambda k_B T/hc = k_B T/\varepsilon = 1/u$, where $u \equiv \varepsilon/k_B T$ is the rescaled energy.
(a) Find the Planck spectral distribution $\mathcal{D}_{ph}(w)$ for a rescaled wavelength such that $dP/P = \mathcal{D}_{ph}(w)\,dw$. (Ignore an overall sign that pops out. We are interested in how much *positive* radiated power dP falls into infinitesimal "bins" of *positive* width du or dw, respectively.)
(b) Use some kind of computer-based tool to plot $\mathcal{D}_{ph}(w)$.
(c) At what wavelength λ_P does your plot of $\mathcal{D}_{ph}(w)$ peak?
(d) Explain why $\lambda_P \neq hc/\varepsilon_P$. (*Hint:* Consider how the "bin sizes" du and dw for the distributions compare as u increases, and recall that only the products $\mathcal{D}_{ph}(u)\,du = \mathcal{D}_{ph}(w)\,dw = dP/P$ have direct physical meaning.)
(e) The sun's surface has a temperature of about 5800 K. What is ε_P for sunlight, and what wavelength is associated with this photon energy?
(f) What is λ_P for light emitted by the sun, and what is the corresponding photon energy?

Rich Context

T6R.1 We can use the Maxwell-Boltzmann distribution to learn some things about planetary atmospheres.
(a) The earth's upper atmosphere has a fairly high temperature, about 1000 K. A molecule in the upper atmosphere can escape if it has a speed that exceeds the earth's escape speed of 11.2 km/s. What is the probability that an N_2 molecule has such a speed?
(b) What is the probability for an H_2 molecule?
(c) Why do you think earth's atmosphere contains nitrogen but not a significant amount of hydrogen?
(d) The escape speed at the moon's surface is 2.4 km/s, where the temperature can exceed 375 K. Why does the moon have no significant atmosphere? (*Hint:* What might be the most massive, reasonably common, naturally occurring molecules in the earth's atmosphere?)

T6R.2 Suppose that a mammal's metabolic rate P (the rate at which it turns food energy into other forms of energy) is primarily determined by the rate at which it emits energy from its surface. Also assume that mammals have nearly the same body temperature T and mass density.
(a) What should be the exponent a in the law $P = Km^a$ that connects a mammal's metabolic rate to its mass m? (K is a constant that we would determine empirically.)
(b) The observed law for mammals seems to be $P = K_0 m^{3/4}$ for $m = 1$ g to 1000 kg (this is **Klieber's law**). Suppose we set the constants in both Klieber's law and the law you found in part (a) so that they agree with the average empirical P for mammals with $m = 1$ kg. About how accurately must we know a 1000-kg mammal's metabolic rate to distinguish between the laws?
(c) Different mammal species with the same body mass have metabolic rates that might differ by even as much

as a factor of five, so comment on how easy it would be to distinguish your law from Klieber's.

Comment: The exact exponent in Klieber's law is still debated, but one reasonably plausible argument based on optimizing the circulatory systems in mammals supports an exponent of 3/4.

T6R.3 The cosmic microwave background (CMB) consists of photons that we now are receiving from distant regions of the universe all around us that were emitted when those regions became transparent about 375,000 years after the Big Bang. This background now looks like blackbody radiation corresponding to a temperature of $T = 2.73$ K.

(a) What is the *average* energy per CMB photon (not the most probable energy)? (*Hint:* See problem T6D.5.)

(b) Roughly how many cosmic microwave photons hit your body every second when you are outdoors? Assume that such photons cannot go through the earth, but can go through the atmosphere. (*Hint:* First imagine yourself floating in space.)

T6R.4 A black hole is about as perfect a blackbody as one can find. Even though a black hole captures all photons falling on it, and photons cannot escape from its interior, quantum processes associated with its event horizon emit photons, called **Hawking radiation.** For a black hole of mass M, this radiation looks exactly like what a blackbody

would emit at a temperature $T = hc^3/16\pi^2 k_B GM$, where G is the universal gravitational constant. A black hole's event horizon has a radius of $R = 2GM/c^2$.

(a) The wavelength λ of a photon with energy ε is $\lambda = hc/\varepsilon$. Compare the wavelength of photons with the most probable energy with the horizon radius R.

(b) Argue that the power P of Hawking radiation that a black hole emits is proportional to M^{-2} and find the constant of proportionality.

(c) The energy for this radiation comes from the black hole's mass energy Mc^2. The emission will therefore eventually cause the black hole to evaporate. Find an expression for how long a black hole of mass M will survive before evaporating. (*Hint:* Express P in terms of $-dM/dt$, then isolate the factors of M on one side and the dt on the other and integrate.)

(d) Before the Large Hadron Collider (LHC) was turned on, some people were concerned that the high energy densities produced by collisions in the detector might create microscopic black holes with mass-energies on the order of 10 TeV. These black holes might then fall into the earth's core, where they would collect and slowly eat up the earth from the inside. This concern is absurd for a host of physical reasons, but one is that such black holes don't survive very long at all. Calculate the farthest that a newly created black hole generated by the LHC might travel before evaporating.

ANSWERS TO EXERCISES

T6X.1 According to equation T6.7b, we have

$$v_P = \sqrt{\frac{2k_B T}{m}} = \sqrt{\frac{2k_B T}{M_A/N_A}}$$

$$= \sqrt{\frac{2(1.38 \times 10^{-23}\, \text{J/K})(295\, \text{K})}{(0.032\, \text{kg})/(6.02 \times 10^{23})}\left(\frac{1\, \text{kg·m}^2/\text{s}^2}{1\, \text{J}}\right)}$$

$$= 391\ \text{m/s} \tag{T6.35}$$

T6X.2 Using the chain rule, we have

$$E_{\text{avg}} = -\frac{1}{Z}\frac{d}{d\beta}\left(\frac{1}{1-e^{-pc\beta}}\right) = -\frac{1}{Z}\left(-\frac{d(-e^{-pc\beta})/d\beta}{[1-e^{-pc\beta}]^2}\right)$$

$$= +\frac{1}{Z}\frac{-e^{-pc\beta}(-pc)}{[1-e^{-pc\beta}]^2} = +[1-e^{-pc\beta}]\frac{pc\,e^{-pc\beta}}{[1-e^{-pc\beta}]^2}$$

$$= \frac{pc\,e^{-pc\beta}}{1-e^{-pc\beta}} = \frac{pc}{e^{pc\beta}-1}, \quad \text{as claimed.} \tag{T6.36}$$

T6X.3 Note that $k_B T = 5(k_B T_{\text{room}}) = 5(0.0254\ \text{eV}) = 0.127\ \text{eV}$.

(a) This means that the most probable photon energy is $\varepsilon_P = 2.82(k_B T) = 0.358\ \text{eV}$.

(b) Note that $u = \varepsilon/k_B T = (1.77\ \text{eV})/(0.127\ \text{eV}) = 13.9$ for the red end of the spectrum and $u = (3.1\ \text{eV})/(0.127\ \text{eV}) = 23.9$ for the violet end of the spectrum. If we substitute these values in as the integral's limits in the Planck web application, we find that

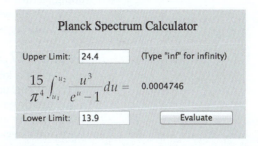

So only about 0.05% of the electromagnetic power emitted by the object will be in the visual portion of the spectrum.

T7

Gas Processes

Chapter Overview

Introduction

In this chapter, we will use the ideal gas model introduced in chapter T5 to explore thermal *processes* in which the properties of a gas change with time. This chapter provides essential background for chapters T8 and T9.

Section T7.1: Work during Expansion or Compression

Because a gas molecule rebounds from a moving piston with a different energy than it had originally, changing a gas's volume involves work, which we can calculate as follows:

$$dW = -P\,dV \tag{T7.4}$$

- **Purpose:** This equation expresses the work dW done on a gas during a volume change dV, where P is the gas pressure.
- **Limitations:** The volume change must be small enough that P does not change significantly during the process; and it must be slow enough that the forces on the piston are essentially in balance, the piston's kinetic energy is negligible, and the gas is in equilibrium with itself.
- **Note:** To find the total work W done on the gas during a process in which the pressure varies, one must integrate this expression: see section T7.4.

Section T7.2: The State of a Gas

A gas's **macroscopic properties** (such as its pressure P, volume V, temperature T, number of molecules N, and thermal energy U) are quantities that we can measure macroscopically (that is, without detailed measurements of what its molecules are doing). It turns out that knowing a suitably chosen *triplet* of such properties (such as P, V, and N) allows us to calculate a gas's other macroscopic properties. Such a triplet describes the gas's macrostate.

Section T7.3: *PV* Diagrams and Constrained Processes

So, when N is known and fixed, knowing P and V determines a gas's macrostate. We can represent such a state by a point on a graph of P versus V (a **PV diagram**). During a **quasistatic gas process** (a process slow enough that even as P changes in time, its value remains uniform throughout the gas), the point representing the gas's state traces out a curve on a PV diagram.

A **constrained process** is a process during which we constrain the gas by holding something constant. During an **isothermal process,** we constrain the gas's temperature to be constant; during an **isobaric process,** we hold its pressure constant; during an **isochoric process,** we keep its volume constant; during an **adiabatic process,** we keep heat energy from flowing into or out of the gas (see figure T7.4). These constrained processes are useful approximations for realistic gas processes.

Section T7.4: Computing the Work

To compute the work W done during a process where P changes significantly, we must evaluate the integral

$$W = -\int P\,dV \qquad \text{(T7.6)}$$

We can evaluate this integral mathematically if we can express P as a function of V, which we can do easily for isochoric, isobaric, and isothermal processes (see equations T7.7, T7.8, and T7.10). But the most important thing to note is that the magnitude of this integral corresponds to the *area under the process's curve on a PV diagram*, an idea we can use to quickly estimate and compare the work involved in various processes.

Section T7.5: Adiabatic Processes

Determining how P, V, and T are related in an adiabatic process is trickier than for the other processes. Using the ideal gas law and the expression for the internal energy of an ideal gas, we find in this section that

$$PV^\gamma = \text{constant} \qquad TV^{\gamma-1} = \text{constant} \qquad \text{(T7.16a,b)}$$

- **Purpose:** These equations specify how a gas's pressure P and temperature T vary with its volume during an adiabatic process. The quantity γ is the gas's **adiabatic index:** $\gamma \equiv 1 + (2/f)$, where f is the number of Newtonian "degrees of freedom" a molecule has for storing energy.
- **Limitations:** The process must be adiabatic (meaning that no heat enters or leaves the gas), the gas must be ideal, and the compression or expansion must occur slowly enough so that P and T have well-defined uniform values throughout the gas during the process.
- **Notes:** For monatomic gases, we have $\gamma = 5/3 = 1.67$; for diatomic gases, $\gamma = 7/5 = 1.40$. For other kinds of (ideal) gases, we can find f empirically by measuring dU/dT and calculating

$$f = \frac{2}{Nk_B}\frac{dU}{dT} \qquad \text{(T7.17)}$$

This result will be useful to us in chapters T8 and T9.

Section T7.6: Ideal Gas Heat Capacities

Because a gas can expand when heated, we must specify whether we are constraining the gas's *volume* to be constant or its *pressure* to be constant when we measure its heat capacity dQ/dT, because the results differ:

$$C_V = \frac{dU}{dT} = \frac{f}{2}Nk_B, \qquad C_P = C_V + Nk_B = \gamma C_V \qquad \text{(T7.24)}$$

- **Purpose:** These equations describe the relationships between a gas's heat capacity C_V measured at constant volume and its heat capacity C_P measured at constant pressure, where N is the number of gas molecules, k_B is Boltzmann's constant, γ is the gas's adiabatic index, and f is the number of Newtonian degrees of freedom for a gas molecule.
- **Limitations:** These relationships strictly apply only to ideal gases, but are good approximations at typical gas densities.

C_P is larger than C_V because heating a gas at constant pressure causes it to expand, which does work pushing back the atmosphere and so requires extra heat to flow in. This is not an issue for most solids or liquids, which expand very little when heated.

T7.1   Work during Expansion or Compression

In chapters T5 and T6, we saw how an understanding of gases at the microscopic level helps us understand how gases behave macroscopically. In this chapter, we will explore the implications of the ideal gas model we developed in chapter T5 to understand more thoroughly how gases behave at the macroscopic level.

Heat and work

In chapter T1, we reviewed the definitions of heat Q and work W (concepts first introduced in unit C). We saw there that both heat and work represent energy *flowing* across a system boundary: heat is an energy flow driven by a temperature difference across that boundary, while work W is an energy flow driven by external forces that act on the system. Work can be done on or by a system in a variety of ways, but in the remainder of the unit we will be primarily concerned with work done as a result of the compression or expansion of a gas. Understanding how work is involved in expansion and compression will be especially important to us in chapter T9, when we discuss how heat can be converted to mechanical energy by using a **heat engine** (such as a gasoline or steam engine). Most heat engines use expansion and compression of some gas to perform this conversion.

Suppose that some external interaction exerts a force $\vec{F}_{ext}$ that pushes a piston through an infinitesimal displacement $d\vec{r}$ that compresses a gas in a cylinder (see figure T7.1). The work that this interaction does on the piston is

$$W = \vec{F}_{ext} \cdot d\vec{r} = |\vec{F}_{ext}||d\vec{r}| \tag{T7.1}$$

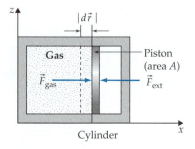

Figure T7.1
Slowly compressing gas in a cylinder.

since $\vec{F}_{ext}$ and $d\vec{r}$ point in the same direction in this case. Now, the gas in the cylinder exerts an opposing force $\vec{F}_{gas}$ against the piston. If we push the piston *slowly* enough that its speed remains essentially constant and its kinetic energy change is insignificant, then the net force on the piston must be essentially zero, so the force $\vec{F}_{gas}$ that the gas exerts on the piston must be essentially equal in *magnitude* to the external force $\vec{F}_{ext}$ on the piston (although they have opposite directions): $|\vec{F}_{ext}| \approx |\vec{F}_{gas}|$. Moreover, $|\vec{F}_{gas}|$ in this case is equal to the pressure P that the gas exerts on the piston times the piston's area A. If we plug these results into equation T7.1, we get

$$dW \approx |\vec{F}_{gas}||d\vec{r}| = PA|d\vec{r}| \tag{T7.2}$$

Now, note that the infinitesimal change in the gas's *volume* in this process is

$$dV \equiv V_{final} - V_{initial} = A(L_{final} - L_{initial}) = -A|d\vec{r}| \quad \text{for compression} \tag{T7.3}$$

This result is negative because the length of the cylinder has *decreased* by $|d\vec{r}|$, meaning that the cylinder's volume has decreased. Substituting this into equation T7.2, we get

Work done by an infinitesimal slow volume change

$$dW = -P\,dV \tag{T7.4}$$

- **Purpose:** This equation expresses the work dW done on a gas during a volume change dV, where P is the gas pressure.
- **Limitations:** The volume change must be small enough that P does not change significantly during the process; and it must be slow enough that the forces on the piston are essentially in balance, the piston's kinetic energy is negligible, and the gas is in equilibrium with itself.
- **Note:** To find the total work W done on the gas during a process in which the pressure varies, one must integrate this: see section T7.4.

(a)   (b)   (c)

Figure T7.2
A ball thrown against the back of a truck will bounce back with (a) about the same energy as it had originally if the truck is at rest, (b) more energy if the truck is backing up, (c) less energy if the truck is moving forward.

The minus sign ensures that dW is positive (energy is flowing into the gas) for a compression even though dV is negative.

Exercise T7X.1

Consider the case of an infinitesimal *expansion*. Go through the steps outlined earlier, and show that even though the signs of various equations along the way are different, you still end up with equation T7.4.

How exactly is energy being transferred to the gas in the case of a compression (or extracted from the gas in the case of an expansion)? We can understand the microscopic processes involved in such an expansion by using a simple analogy. Imagine bouncing tennis balls off the back end of a truck (see figure T7.2). If the truck is stationary and the balls are perfectly elastic, they will bounce back with about the same kinetic energy as they had before striking the truck. But if the truck is backing up toward you, the balls will bounce back from the truck with more energy relative to you than you gave them: some of the energy of the truck's motion is being converted to kinetic energy in the balls. Similarly, if the truck is moving away from you, the balls will bounce back with less energy than they had to begin with: energy is being transferred from the balls to the truck.

Similarly, gas molecules that bounce back elastically from a piston at rest bounce back from an advancing piston (during compression) with more energy than they had originally, implying that the gas's thermal energy increases at the expense of the piston's energy. When the piston retreats (during expansion), the molecules bounce back with less energy than they had originally and so transfer the gas's thermal energy to the piston.

A microscopic model for how energy is transformed

T7.2   The State of a Gas

What properties of a sample of gas can we measure at the macroscopic level—that is, without measuring properties of the sample's individual molecules? We can certainly measure the gas sample's pressure P, its volume V, and its temperature T without even knowing that molecules exist. We can also measure its total internal thermal energy U (at least in principle) by measuring how much energy we have to add to the gas to increase its temperature from absolute zero to its current temperature (while keeping its volume fixed). Finally, we can measure the gas sample's mass M, and if we know the gas's chemical composition, we can use this to calculate the number of moles and (given a value for Avogadro's number N_A) the number of molecules N in the sample. The quantities P, V, T, U, M, and N are, therefore, some of the gas's macroscopically variable properties.

Macroscopically variable properties of a gas

Note that the six quantities P, V, T, U, M, and N are *not* all independent. In chapter T5, we saw that these quantities are linked by the three equations

$$N = \frac{M}{M_A} N_A, \qquad PV = N k_B T, \qquad \text{and} \qquad U = \frac{f}{2} N k_B T \qquad (T7.5)$$

(Note that if we know the gas's chemical composition, in principle we know its molar mass M_A and the number of energy-storage "degrees of freedom" f.) Therefore, if we know the sample's chemical composition and a suitably chosen *triplet* of properties, such as P, V, N or U, V, N, we can use these three equations to calculate the sample's other three macroscopic properties.

The macrostate of a gas

This means that two gas samples having the same chemical composition and the same values for a suitably chosen triplet of values will be physically indistinguishable if we are not allowed to examine the behavior of individual molecules. We say, therefore, that such a triplet of values describes the macrostate of an ideal gas.

T7.3   *PV* Diagrams and Constrained Processes

Representing states and processes on a *PV* diagram

In many situations of interest, the number of molecules N in a sample of gas is known and fixed. When this is true, we need to know only two additional gas properties to completely determine its macrostate at a given instant of time. If we choose these variables to be P and V, then every macrostate of a fixed amount of gas corresponds to a unique *point* on a graph of P versus V, as shown in figure T7.3. We call such a graph a *PV* **diagram.**

Since $PV = N k_B T$, all the points on a *PV* diagram corresponding to a given temperature lie on a curve such that $PV = $ constant, or $P \propto 1/V$. A set of such curves corresponding to various values of T is also shown in figure T7.3.

A **quasistatic gas process** is any process that changes the state of a gas and that is slow enough that the gas remains essentially in equilibrium at all times. This means that at all times the gas pressure P has the *same* well-defined value at all points in the gas. As the state of the gas changes, the point representing that state on the *PV* diagram will move, marking out a *path* on the diagram. An example path is also shown in figure T7.3.

Important (and useful) gas processes

P and V are completely independent variables, and there is no intrinsic reason why a gas cannot have any pressure at any given volume. A process in principle (like the "general process" shown in the figure) might involve any connected sequence of points on the diagram. In practice, though, there are four special gas processes that keep coming up as useful approximations in realistic situations:

Figure T7.3

Every point on this *PV* diagram represents a possible macroscopic state for a fixed amount of ideal gas. The thin curves connect gas states having the same temperature (the curves are labeled assuming that $N = 5.8 \times 10^{23}$). Any quasistatic process is an ordered succession of states and can be represented by a succession of points (that is, a curve) on this diagram.

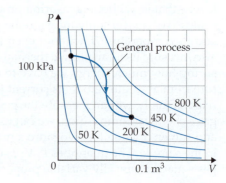

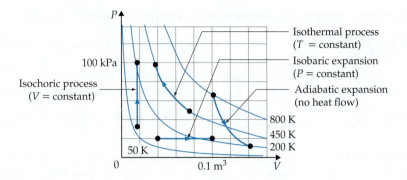

Curves on a *PV* diagram corresponding to the four most useful types of processes. Again, the temperature labels on the isothermal curves assume that $N = 5.8 \times 10^{23}$.

1. In **isochoric processes,** the gas is heated or cooled while its volume is constrained to be constant (for example, by keeping the gas in a rigid container). The root *iso-* means "same," and *choric* refers to volume.
2. In **isobaric processes,** the gas is heated or cooled while its *pressure* is constrained to be constant (for example, by confining the gas with a piston whose other side is acted on by the constant pressure of the earth's atmosphere). The root *bar-* refers to pressure (as in the word *barometer*).
3. In **isothermal processes,** the gas is expanded or compressed while its temperature is constrained to be constant (for example, by placing it in good thermal contact with a "bath" having a certain fixed temperature).
4. In **adiabatic processes,** the gas is expanded or compressed while heat flow to or from the gas is constrained to be zero (for example, by putting thermal insulation around the gas container). We will discuss adiabatic processes more fully in section T7.5.

Figure T7.4 illustrates these processes. We call them **constrained processes,** because the path on the *PV* diagram that each process follows is determined by a *constraint* placed on the gas during the process (for example, the constraint that the gas's volume, pressure, or temperature remains fixed).

T7.4 Computing the Work

These constrained processes are useful partly because we can actually compute the work done on or by the gas involved in such a process. As we noted before, the equation $dW = -P\,dV$ applies only to *infinitesimal* compressions or expansions, because as a gas is compressed or expanded by a finite amount, its pressure P will generally change during the process. If we want to compute the total work W done during a volume change during which P changes significantly, we must evaluate the integral

Calculating W when the pressure changes significantly during a process

$$W = -\int P\,dV \tag{T7.6}$$

In words, this equation tells us that to compute the total work, we should divide the process into infinitesimal steps, compute $dW = -P\,dV$ for each step, and sum the result over all the steps.

To actually evaluate such an integral, we need to be able to express P during the process as a function of V. This is easy to do for each of the four special constrained processes listed in section T7.3 except for the adiabatic process, which we will consider in the next section. The *isochoric* process (where V is constant) is trivial: since the volume does not change ($dV = 0$), no work is done as the result of compression or expansion:

Implications for particular constrained processes

$$W = 0 \quad \text{(isochoric)} \tag{T7.7}$$

In an *isobaric* process, P is constant, so we can pull it out of the integral:

$$W = -\int P\,dV = -P\int_{V_i}^{V_f} dV = -P(V_f - V_i) = -P\Delta V \quad \text{(isobaric)} \qquad \text{(T7.8)}$$

where V_i is the gas's initial volume and V_f is its final volume.

In an isothermal process, T is constant. The ideal gas law then implies

$$PV = Nk_B T = \text{constant} = P_i V_i \quad \Rightarrow \quad P(V) = \frac{Nk_B T}{V} = \frac{P_i V_i}{V} \qquad \text{(T7.9)}$$

where P_i and V_i are the initial pressure and volume of the gas, respectively, and $P(V)$ means "pressure as a function of volume." If you substitute this into equation T7.6, and use $\int x^{-1} dx = \ln x$ and $\ln x - \ln y = \ln(x/y)$, you can show that

$$W = -Nk_B T \ln\frac{V_f}{V_i} = -P_i V_i \ln\frac{V_f}{V_i} \quad \text{(isothermal)} \qquad \text{(T7.10)}$$

Exercise T7X.2

Verify that equation T7.10 is correct.

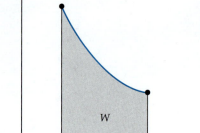

Figure T7.5

The *magnitude* of the work flowing into or out of a gas is equal to the area under the curve of P when plotted as a function of V. The sign of the work is positive if the gas is being compressed, negative if it expands.

One of the reasons that PV diagrams of gas processes are so useful is that *the work done in a given quasistatic expansion or compression process is equal in magnitude to the area under the curve representing that process on a PV diagram.* This is a direct consequence of equation T7.6: if we consider P to be a function of V, then the standard interpretation of the integral of $P(V)$ is that it corresponds to the area of the curve of P when it is plotted as a function of V (see figure T7.5).

Because of this simple visual interpretation of equation T7.6, one can get a lot of qualitative information about a process from a PV diagram, as the following example demonstrates:

Example T7.1

Problem: Suppose we take an ideal monatomic gas from its initial state A to state B by an *isothermal* process, from B to C by an *isobaric* process, and from C back to its initial state A by an *isochoric* process, as shown in figure T7.6a. Determine the signs of Q, W, and ΔU (or state that the value is zero) for each step (assuming that no energy flows into or out of the gas *except* in the form of heat or from work due to expansion or compression).

Solution Figure T7.6b shows a chart of the results. which we can determine as follows. Process $A \rightarrow B$ is an isothermal expansion, meaning that T is constant. But for an ideal gas, U is proportional to T (because $U = [f/2]Nk_B T$) so ΔU is *zero* during any isothermal process. During an expansion, work energy flows *out* of the gas (for each infinitesimal step along this process, $dW = -P\,dV$, and $dV > 0$ in an expansion process), so W is *negative*. But if work energy flows *out* of the gas and yet the gas's total thermal energy U remains the same, then (if no other kind of energy flows across the boundary), then heat energy must flow *into* the gas, meaning that Q must be *positive*.

Process $B \rightarrow C$ is an isobaric *compression*, so W is *positive* here. On the other hand, the ideal gas law says that $PV = Nk_B T$. Therefore, since V is decreasing in this process while P remains constant, T must be decreasing (that is, $dT < 0$ for each small step in the process) which in turn implies that the thermal energy U of the gas must be decreasing: thus ΔU is *negative*. Since

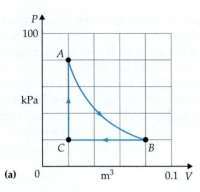

(a)

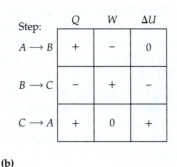

Step:	Q	W	ΔU
$A \rightarrow B$	$+$	$-$	0
$B \rightarrow C$	$-$	$+$	$-$
$C \rightarrow A$	$+$	0	$+$

(b)

Figure T7.6
A sequence of gas processes.

work energy is flowing *into* the gas and yet its thermal energy is *decreasing*, heat energy must be flowing *out* of the gas: Q is negative.

Finally, process $C \rightarrow A$ is an isochoric process. Since there is no change in volume, no work is done on or by the gas: $W = 0$. Yet the temperature is increasing (since PV is increasing), so the gas's thermal energy is increasing, meaning that ΔU is *positive* in this process. This increase in thermal energy must be supplied by heat (since $W = 0$), so Q is positive.

We can also estimate, directly from the diagram, the work energy flowing into or out of the gas in the cyclic process we've been considering. Each grid square's worth of area on the PV diagram shown in figure T7.7 represents

Estimating W from a PV diagram

$$(20 \times 10^3\,\cancel{Pa})(0.020\,\cancel{m^3})\left(\frac{1\,\cancel{N/m^2}}{1\,\cancel{Pa}}\right)\left(\frac{1\,J}{1\,\cancel{N \cdot m}}\right) = 400\,J \qquad (T7.11)$$

There are a total of about six squares (four whole squares, two squares mostly complete, and two small parts of squares) of area under the curve for the isothermal expansion $A \rightarrow B$, so the gas loses $6 \times 400\,J = 2400\,J$ of work energy during that process. In the isobaric compression $B \rightarrow C$, the gas gains $3 \times 400\,J = 1200\,J$ of work energy. Since $W = 0$ for the isochoric process, the total energy lost by the gas during the entire cyclic process is about 1200 J. Note that *the net work done by the gas in a cyclic process is equal to the area enclosed by the process*: this is a useful general statement.

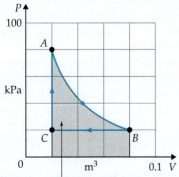

(a) Work energy flowing *out* of the gas during the isothermal process $A \rightarrow B$.

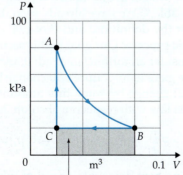

(b) Work energy flowing *into* the gas during the isobaric process $B \rightarrow C$.

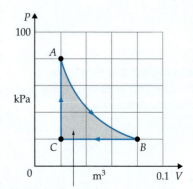

(c) *Net* work energy flowing *out* of the gas during the entire cyclic process.

Figure T7.7
The net work energy flow during a cyclic process.

Processes that convert heat to work

Note that since the gas comes back to the same state *A* (and thus the same temperature) at the end of the process that it had originally, its thermal energy must be the same at the end of the cycle as it was in the beginning (since *U* for an ideal gas only depends on *N* and *T*). Yet the gas loses about 1200 J of work energy in the cycle, as we have just seen. Where does this energy come from, if not from the gas's thermal energy? It must in fact come from the *heat* energy flowing into the gas. This cycle is therefore an example of a process that converts heat energy to work energy. Many kinds of heat engines use such gas processes. (We will talk more about heat engines in chapter T9.)

Exercise T7X.3

The figure below shows another cyclic gas process. Fill out the chart as we did in example T7.1. Also estimate the net amount of work energy flowing *out* of the gas, and indicate the step in which heat flowed *into* the gas.

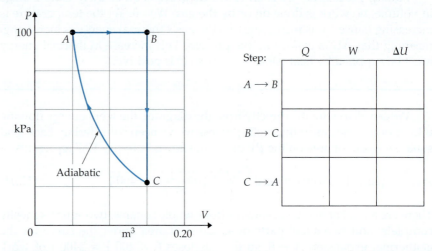

T7.5   Adiabatic Processes

Exercise T7X.3 included an *adiabatic* compression process as one of the steps in the cyclic process. In an adiabatic process, no heat is permitted to enter or leave the system in question. (*Adiabatic* comes from Greek *adiabatos*, meaning "impassable.") We have already seen that during an isothermal process, *T* is constant (by definition) and $P \propto 1/V$. How do *T* and *P* depend on *V* in an adiabatic process? Our goal in this section is to find out.

Deriving the law that links pressure and volume during adiabatic processes

Consider an infinitesimal adiabatic volume change dV. During this process, the gas's temperature will change by some tiny amount dT and the pressure will change by some tiny amount dP. If we take the derivative of the ideal gas law with respect to absolute temperature *T*, we get

$$P\frac{dV}{dT} + V\frac{dP}{dT} = Nk_B \quad \Rightarrow \quad P\,dV + V\,dP = Nk_B\,dT \tag{T7.12}$$

Now if no heat flows into or out of this gas in this process, then the change in the gas's internal energy *U* must be entirely due to work. But because the change in an ideal gas's internal energy is $dU = \frac{1}{2}fNk_B\,dT$, we have

$$-P\,dV = dW = \frac{f}{2}Nk_B\,dT \tag{T7.13}$$

where f is the number of the molecules' "degrees of freedom." (Note that for polyatomic gases f might not be an integer, as it is for monatomic and diatomic gases, but the *value* of f does not matter here as long as dU depends *only* on dT.) We can use this to eliminate $Nk_B\,dT$ from equation T7.12:

$$P\,dV + V\,dP = -\frac{2}{f}P\,dV \quad\Rightarrow\quad \gamma P\,dV + V\,dP = 0 \tag{T7.14}$$

where we call $\gamma \equiv 1 + 2/f$ the gas's **adiabatic index.** (Note that for a monatomic gas, $\gamma = 1 + 2/3 = 5/3$; for a diatomic gas $\gamma = 1 + 2/5 = 7/5$.) Now if we multiply both sides of equation T7.14 by $V^{\gamma-1}/dV$ we see that

$$\gamma PV^{\gamma-1} + V^\gamma \frac{dP}{dV} = 0 \quad\Rightarrow\quad \frac{d}{dV}(PV^\gamma) = 0 \tag{T7.15}$$

(Check this by using the product rule to evaluate the derivative in the right-hand equation to see that you get the left-hand equation.) This means that in any adiabatic process, PV^γ *must be a constant.* So, where we had $PV = $ constant for isothermal processes, we have $PV^\gamma = $ constant for adiabatic processes.

If you want to know how the gas's temperature varies during the process, you can apply the ideal gas law to show that $PV^\gamma = $ constant implies that $TV^{\gamma-1}$ is a constant (see problem T7D.1). So, to summarize:

$$PV^\gamma = \text{constant} \tag{T7.16a}$$

$$TV^{\gamma-1} = \text{constant} \tag{T7.16b}$$

Laws connecting P, V, and T for adiabatic processes

- **Purpose:** These equations specify how a gas's pressure P and temperature T vary with its volume during an adiabatic process. The quantity γ is the gas's adiabatic index: $\gamma \equiv 1 + (2/f)$, where f is the number of Newtonian "degrees of freedom" a molecule has for storing energy.
- **Limitations:** The process must be adiabatic (meaning that no heat enters or leaves the gas), the gas must be ideal, and the compression or expansion must occur slowly enough so that P and T have well-defined uniform values throughout the gas during the process.
- **Notes:** For monatomic gases, we have $\gamma = 5/3 = 1.67$; for diatomic gases, $\gamma = 7/5 = 1.40$. For other kinds of (ideal) gases, we can find f empirically by measuring dU/dT and calculating

$$f = \frac{2}{Nk_B}\frac{dU}{dT} \tag{T7.17}$$

Note that in *applying* equations T7.16, initial conditions determine the value of the constant. For example, at point C in the adiabatic process shown in exercise T7X.3, the constant value of PV^γ throughout the process to point A must be the same as the value of $P_C V_C^\gamma$. To actually *calculate* this value, we need to know the value of γ, but we can actually get this from the graph! Since we must have $P_A V_A^\gamma = P_C V_C^\gamma$, and since P increases by a factor of roughly $100\text{ kPa}/22\text{ kPa} = 4.55$ when V decreases by a factor of 3, we see that

How to find the value of the constant

How to find the value of γ

$$1 = \frac{P_A V_A^\gamma}{P_C V_C^\gamma} = \left(\frac{P_A}{P_C}\right)\left(\frac{V_A}{V_C}\right)^\gamma \approx 4.55\left(\frac{1}{3}\right)^\gamma \quad\Rightarrow\quad 0 \approx \ln 4.55 - \gamma \ln 3$$

$$\Rightarrow\quad \gamma \approx \frac{\ln 4.55}{\ln 3} = 1.38 \tag{T7.18}$$

So the gas in this example is probably diatomic.

Example T7.2

Problem: Suppose we compress a sample of air whose initial pressure is 100 kPa and temperature is 22°C (= 295 K) to a volume that is one-quarter of its original volume within 0.1 second (s). The piston in this case moves a total distance of 30 cm. Is this compression too fast to use the adiabatic model? If not, what is the gas's final temperature?

Solution   In general, a compression or expansion will be too fast if the piston's speed is a significant fraction of the gas molecules' thermal speeds: in such a case, the gas molecules near the piston would not have enough time to communicate what the piston is doing to the rest of the gas during the process, meaning that the gas's pressure and temperature could be significantly different near the piston than elsewhere in the gas. But in this case, the piston's average speed is about 3 m/s. This is much smaller than the typical thermal speeds of gas molecules, which are on the order of hundreds of meters per second at room temperature. So this process is probably slow enough. Indeed, the advantage of such a rapid process is that, even if the cylinder is not very well insulated, everything happens so fast that not much heat will be able to flow into or out of the cylinder. Therefore, we can quite reasonably model this process as an adiabatic compression. Since air is a diatomic gas and is reasonably ideal at normal pressures and temperatures, it should obey the equation $TV^{\gamma-1}$ = constant with $\gamma = 1.4$. In this case, therefore,

$$T_f V_f^{\gamma-1} = T_i V_i^{\gamma-1} \;\Rightarrow\; T_f = T_i \left(\frac{V_i}{V_f}\right)^{\gamma-1} = 295\ \text{K} \left(\frac{V_i}{\frac{1}{4}V_i}\right)^{0.40} = 514\ \text{K} \qquad (T7.19)$$

Two methods for calculating the work involved in an adiabatic process

Now that we know how P varies with V during an adiabatic process, we can calculate the work W done during such a process. Suppose that our gas initially has pressure P_i and volume V_i, and we compress or expand it adiabatically to a final volume V_f. Because $PV^\gamma = P_i V_i^\gamma$ at all times during the process, the gas pressure as a function of volume is therefore $P = P_i V_i^\gamma V^{-\gamma}$. You can then show (see problem T7D.2) that the work done is

$$W = -\int_{V_i}^{V_f} \left(\frac{P_i V_i^\gamma}{V^\gamma}\right) dV = \frac{P_i V_i}{\gamma-1}\left[\left(\frac{V_i}{V_f}\right)^{\gamma-1} - 1\right] \qquad (T7.20)$$

This formula is not as pretty as the isothermal formula, we will find it useful for understanding certain kinds of heat engines in chapter T9.

However, when using PV diagrams, one can usually more easily estimate the work done in an adiabatic process by calculating Nk_BT at the beginning and the end of the process using $Nk_BT = PV$, and then noting that since no heat flows in an adiabatic process, $W = \Delta U = (f/2)Nk_B\Delta T = (f/2)\Delta(PV)$. So, for example, if a diatomic gas's pressure goes from 22 kPa to 100 kPa as the volume decreases from 0.15 m³ to 0.05 m³, the work done on the gas to compress it must be $\frac{5}{2}(100\cdot0.05 - 22\cdot0.15)$ kPa·m³ = 4.3 kPa·m³ = 4.3 kJ.

T7.6   Ideal Gas Heat Capacities

The definition of heat capacity

We define a substance's "heat capacity" in practical circumstances to be $C \equiv dQ/dT$, which specifies how much heat dQ one must add to the substance to get a given (tiny) change in temperature dT. With solids, performing such measurements yield reasonably consistent results, but with gases, one does *not* get consistent results without specifying how the gas's *volume* is constrained to behave as we add the heat.

If we constrain the gas's volume to be *constant* during the process, then no work is done on or by the gas as heat is added. The heat added therefore entirely goes to the gas's thermal energy ($dQ = dU$) implying that

$$C = \frac{dQ}{dT} = \frac{dU}{dT} = \frac{f}{2}Nk_B \qquad \text{(at constant volume } V\text{)} \qquad \text{(T7.21)}$$

The heat capacity of a gas measured at constant volume

However, in many practical situations on the earth's surface, a gas has a constant *pressure* P (equal to that of the earth's atmosphere) but its volume is *not* fixed. For example, if we add heat to air in a flexible container surrounded by the earth's atmosphere, the gas's pressure is constant but not its volume. In such a case, adding heat to a gas means that the gas expands and thus does some work on the earth's atmosphere. This means that we must add *more* heat to the gas to get a given temperature increase dT, because some of the heat energy we add to the gas will go to pushing back the atmosphere instead of increasing the gas's thermal energy and thus its temperature. The ideal gas law tells us that when the gas's pressure is held constant, then

The heat capacity of a gas measured at constant pressure

$$dW = -P\,dV = -Nk_B\,dT \qquad \text{(T7.22)}$$

Since in this case $dU = dQ + dW$, the gas's measured heat capacity will be

$$C = \frac{dQ}{dT} = \frac{dU}{dT} - \frac{dW}{dt} = \frac{f}{2}Nk_B + Nk_B = \left(1 + \frac{f}{2}\right)Nk_B \quad \text{(at constant } P\text{)} \quad \text{(T7.23)}$$

We see that the heat capacities in these two situations are significantly different. We call the heat capacity measured the first way the gas's constant-volume heat capacity C_V, and the heat capacity measured the second way the gas's constant-pressure heat capacity C_P. Then, we have

$$C_V = \frac{dU}{dT} = \frac{f}{2}Nk_B, \qquad C_P = C_V + Nk_B \qquad \text{(T7.24}a\text{)}$$

Relationships involving ideal gas heat capacities

$$C_P = \left(\frac{2}{f} + 1\right)\frac{f}{2}Nk_B = \gamma\frac{f}{2}Nk_B = \gamma C_V \qquad \text{(T7.24}b\text{)}$$

- **Purpose:** These equations describe the relationships between a gas's heat capacity C_V measured at constant volume and its heat capacity C_P measured at constant pressure, where N is the number of gas molecules, k_B is Boltzmann's constant, γ is the gas's adiabatic index, and f is the number of Newtonian degrees of freedom for a gas molecule.
- **Limitations:** These relationships strictly apply only to ideal gases, but are good approximations at typical gas densities.

People almost universally measure the heat capacities of solids in the earth's atmosphere *without* constraining the solid's volume to be constant (which would be much harder to do than for a gas). However, a solid's volume typically varies only quite minutely with increasing temperature, so the solid does almost no work on the atmosphere as its temperature increases, meaning that $C_P \approx C_V$. (This is why we did not worry about this issue when we defined $C = dU/dT$ when discussing solids earlier in this volume.) $C_P \approx C_V$ for most liquids as well. But the distinction between C_P and C_V is very important when considering gases.

The distinction is less important for solids and liquids

TWO-MINUTE PROBLEMS

T7T.1   A gas with a pressure of 100 kPa is in a container that is a cube measuring 10 cm on a side. If we move one wall in 1 millimeter (mm), does work flow into (A) or out of (B) the gas? What is the magnitude of W?
A. 1000 J
B. 100 J
C. 10 J
D. 1 J
E. 0.001 J
F. Other (specify)

T7T.2   Which of the following triplets of macroscopic variables would *not* adequately specify a gas's macrostate?
A. U, V, T
B. P, T, N
C. U, N, T
D. M, V, T
E. M, P, V
F. M, P, U

Problems T7T.3 through T7T.7 refer to the following cyclic gas process:

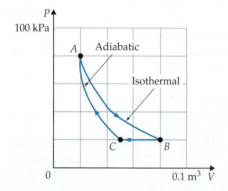

T7T.3   The process $B \rightarrow C$ shown is
A. an isochoric process.
B. an isothermal process.
C. an isobaric process.
D. an adiabatic process.
E. an isometric process.
F. none of the above.

T7T.4   What are the signs of Q, W, and ΔU for the process $A \rightarrow B$?
A. $0, -, -$
B. $0, +, +$
C. $+, -, 0$
D. $+, +, 0$
E. $-, +, 0$
F. Other (specify)

T7T.5   What are the signs of Q, W, and ΔU for the process $B \rightarrow C$? (Select from the answers listed in problem T7T.4.)

T7T.6   What are the signs of Q, W, and ΔU for the process $C \rightarrow A$? (Select from the answers listed in problem T7T.4.)

T7T.7   Is the work energy flowing into or out of the gas in process $B \rightarrow C$ positive (A) or negative (B)? Which of the values below is closest to the magnitude of W?
A. 0.6 J
B. 1.5 J
C. 300 J
D. 600 J
E. 1500 J
F. 3000 J

T7T.8   Suppose that a bubble of helium (a monatomic gas) rising from the bottom of the ocean expands in volume by a factor of 8 by the time it reaches the surface [where the pressure is 1 atmosphere (atm)]. Assume that the bubble rises so fast that it expands essentially adiabatically. What was the pressure on the gas at the depth where it formed (in atmospheres)?
A. 3.5 atm
B. 8 atm
C. 16 atm
D. 18 atm
E. 32 atm
F. Other (specify)

T7T.9   If the temperature of the bubble described in problem T7T.8 was 320 K when it formed, what is its approximate final temperature when it reaches the surface?
A. 40 K
B. 80 K
C. 320 K
D. 1280 K
E. 2560 K
F. Other (specify)

T7T.10   During an adiabatic compression, does a gas's temperature increase, decrease, or remain the same?
A. It increases.
B. It decreases.
C. It remains the same.

T7T.11   Does the pressure of a gas decrease more rapidly or less rapidly when expanding adiabatically than it would when expanding isothermally?
A. More rapidly.
B. Less rapidly.
C. The pressure decreases at the same rate in both cases.
D. The pressure *increases* in an adiabatic expansion.
E. The pressure *increases* in an isothermal expansion.

T7T.12   Water actually contracts when it is warmed near its freezing point. In this range, is water's heat capacity at constant volume (A) greater than, (B) less than, or (C) equal to its heat capacity at constant pressure?

HOMEWORK PROBLEMS

Basic Skills

T7B.1   Suppose I tell you that hydrogen in a rigid container at room temperature has a volume of $V = 0.020$ m^3 and a mass $M = 1.0$ g.
(a) Calculate the number of molecules N in the container.
(b) Calculate the gas's thermal energy U.
(c) Calculate the gas's pressure P.

T7B.2   Suppose we use a piston to confine a gas in a cylinder. The gas has an initial pressure of 120 kPa and a volume of 100 cm^3. We slowly move the piston back until the gas's volume has increased by 0.5%. What is the approximate work that flows into or out of the gas? (Be sure to give the correct magnitude and sign. You may consider this volume change to be small enough that $P \approx$ constant.)

T7B.3   Suppose we use a piston to confine a gas in a cylinder. The gas has an initial pressure of 95 kPa and a volume of 300 cm^3. We slowly push the piston in until the gas's volume has decreased by 1%. What is the approximate work that flows into or out of the gas? (Be sure to give the correct sign, as well as the correct magnitude. You can consider this volume change to be small enough that P is approximately constant during the process.)

T7B.4   Suppose we allow an ideal gas in a cylinder to expand while holding its temperature constant. Does any heat flow in this process? If so, does it flow into or out of the gas? Explain.

T7B.5   Suppose we compress an ideal gas in a cylinder while holding its pressure fixed. Does any heat flow in this process? If so, does it flow into or out of the gas? Explain.

T7B.6   Use equation T7.10 to check the estimate that I made of the work done in process $A \rightarrow B$ in figure T7.7 (see below equation T7.11).

T7B.7   A gas initially at atmospheric pressure (100 kPa) in a box 10 cm on a side is isothermally compressed to one-half its original volume. What is W for this process?

T7B.8   Consider a cyclic gas process that looks like a rectangle on a PV diagram, operating between pressures P_1 and $P_2 > P_1$ and volumes V_1 and $V_2 > V_1$. If the gas goes counterclockwise around the cycle, what is the net work energy flowing out of the gas in this cycle? Explain (perhaps with the help of a sketch).

T7B.9   Suppose we use a piston to confine an ideal gas with an initial pressure of 120 kPa in a cylinder with a volume of 150 cm^3. We then allow the gas to slowly expand to a volume of 350 cm^3 while adding enough heat to keep its pressure fixed. What is the work that flows into or out of the gas? (Specify the correct sign, as well as the amount.)

T7B.10   Suppose we confine a monatomic ideal gas with an initial pressure of 80 kPa in a cylinder with an initial volume of 600 cm^3. We then compress the gas isothermally until its volume has decreased to 450 cm^3. What is its pressure now?

T7B.11   Suppose we confine a monatomic ideal gas with an initial pressure of 60 kPa in a cylinder with an initial volume of 600 cm^3. We then compress the gas adiabatically until its volume has decreased to 450 cm^3. What is its pressure now?

T7B.12   Explain *physically* (no equations!) why a gas's temperature increases as we compress it adiabatically.

Modeling

T7M.1   By exerting a pressure of about 200 atm (20 MPa) one can compress water to 99% of its original volume (that is, compress it by 1%). Suppose that the water's original volume is $V_0 = 1$ liter, and that over such a small range of volume change, P increases proportionally with $V_0 - V$.
(a) Draw this process on a PV diagram.
(b) Estimate the work W that you must do on the water.
(c) Do you find your result surprising? Comment.

T7M.2   Suppose we allow one mole of helium gas in a cylinder to expand adiabatically. During the expansion, the helium's temperature falls from 310 K to 265 K. How much work energy flows in this expansion? (Be sure to give both the correct magnitude and the correct sign.)

T7M.3   Suppose we constrain 3.0×10^{22} molecules of nitrogen gas at 280 K to expand isothermally to 3 times the gas's original volume. Heat must enter the gas during this process. Why and how much?

T7M.4   Suppose we constrain a gas to follow the three-step cyclic process shown in the graph below. Prepare a chart (like the one shown in example T7.1) that specifies the sign of Q, W, and ΔU for each step in the process. What is the net work flowing into or out of the gas for the entire cycle? (Be sure to give the correct magnitude and sign.)

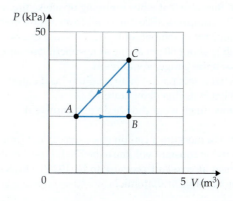

T7M.5 Suppose we constrain a gas to follow the three-step cyclic process shown in the graph below. Prepare a chart (like the one shown in example T7.1) that specifies the sign of Q, W, and ΔU for each step in the process.

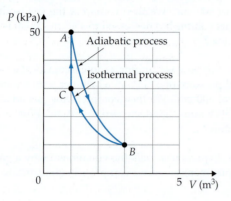

T7M.6 Heat must flow into the gas during the process $C \rightarrow A$ shown in the drawing associated with problem T7M.5. Why? If the gas's temperature is 290 K at point C, find its temperature at point A and the heat that has flowed into the gas in this process if the gas is monatomic.

T7M.7 A bubble of air forms at the bottom of the ocean floor 66 ft below the surface, where the ambient pressure is about 300 kPa = 3 atm. The bubble has an initial volume of about 25 cm^3 and a temperature of 8°C. If the bubble rises to the surface so fast that it expands essentially adiabatically, what is its final volume and temperature?

T7M.8 A research balloon bound for the stratosphere is filled at sea level with 800 m^3 of helium whose initial temperature is 285 K. The balloon is released, and it climbs to an altitude where the air pressure is 0.045 times its value at sea level.
(a) If the helium expands adiabatically, what is the balloon's volume at its final altitude?
(b) What is the helium's final temperature?

T7M.9 An adiabatic expansion process begins with 0.1 mole of gas at a pressure of 100 kPa and ends with a pressure of 40 kPa, while its temperature falls from 300 K to 208 K.
(a) Is the gas diatomic or monatomic?
(b) How much work did the gas do in this process?

T7M.10 Suppose that while pumping up a bike tire, we fairly rapidly compress 1500 cm^3 of air from atmospheric pressure and room temperature to a pressure of about 5 atm (which is about 60 psi above atmospheric pressure, which is what a tire gauge would read).
(a) What is this packet of air's volume as it enters the tire?
(b) What is its final temperature?
(c) How much work did we do to compress it?

T7M.11 We measure a certain gas's *specific heat* (not its heat capacity) at constant volume to be 730 J/(kg·K) and its specific heat at constant pressure to be 1020 J/(kg·K). Is this gas monatomic or diatomic?

T7M.12 Water has a specific heat of 4186 J/kg.
(a) Calculate the effective number of "degrees of freedom" f for water (see equation T7.17).
(b) Speculate physically about why this number might be as large as it is.

Derivation

T7D.1 Use the ideal gas law to show that if PV^γ = constant for an adiabatic process, then $TV^{\gamma-1}$ = constant for such a process.

T7D.2 Use $PV^\gamma = P_iV_i^\gamma$ to show that the work done during an adiabatic volume change from V_i to V_f is

$$W = \frac{P_iV_i}{\gamma - 1}\left[\left(\frac{V_i}{V_f}\right)^{\gamma-1} - 1\right] \qquad (T7.25)$$

(*Hint:* Solve $PV^\gamma = P_iV_i^\gamma$ for P as a function of V, and then use equation T7.6.)

T7D.3 One can find the work involved in an adiabatic process in one of two ways. The first way is to use equation T7.25. The second way is to realize that since no heat flows into the gas, the work that flows into (or out of) the gas is the same as the gas's change in internal energy: $W = \Delta U = Nk_B\Delta T$. Prove mathematically that this second method yields the same result as equation T7.25.

T7D.4 How does T vary with P in an adiabatic process?

T7D.5 One can also derive the adiabatic gas law from a molecular-level perspective. This problem outlines how. Suppose a piston with area A confines a gas to a cylinder of length L, and suppose the piston moves slowly inward with a speed $|\vec{u}| = -dL/dt$ (negative because L is decreasing). Let's define our coordinate system so that the piston moves in the $-x$ direction.
(a) If the piston were at rest, a molecule hitting it would simply reverse its x-velocity. But if the piston is moving inward, argue that the absolute value of the molecule's x-velocity *increases* by $2|\vec{u}|$. (*Hint:* One can do this either by transforming to the frame in which the piston is at rest and then back again, or by solving the one-dimensional elastic collision problem for a light object hitting a much more massive moving object.)
(b) Show that this means that the kinetic energy of a molecule with x-velocity v_x increases by $dK \approx 2m|\vec{u}||v_x|$ per collision in the limit that $|\vec{u}| \ll |v_x|$.
(c) Assuming that the molecule does not interact with other molecules, the time between collisions with the piston will be $2L/|v_x|$. Show then that the rate at which the molecule's energy will increase with time is

$$\frac{dK}{dt} = \frac{m|\vec{u}||v_x|^2}{L} \qquad (T7.26)$$

(d) The total amount of energy that the gas gains from the piston is therefore the number of molecules N times the value of the right side of equation T7.26 averaged over all molecules. Argue that this will be

$$\frac{dU}{dt} = \frac{N|\vec{u}|}{L} k_B T \qquad (T7.27)$$

(*Hint:* See equation T5.27.)

(e) For an ideal gas, $U = \frac{1}{2}fNk_BT \Rightarrow dU/dt = \frac{1}{2}fNk_B\,dT/dt$, where f is the number of degrees of freedom for a molecule. Substitute this into equation T7.27 to get

$$\frac{1}{T}\frac{dT}{dt} \approx \frac{2}{f}\frac{|\vec{u}|}{L} = (\gamma - 1)\frac{|\vec{u}|}{L} \qquad (T7.28)$$

(f) Argue that

$$\frac{|\vec{u}|}{L} = -\frac{1}{V}\frac{dV}{dt} \qquad (T7.29)$$

(g) Substitute this into equation T7.28, multiply both sides by $TV^{\gamma-1}$, and from what you get, argue that

$$\frac{d}{dt}(TV^{\gamma-1}) = 0 \quad \Rightarrow \quad TV^{\gamma-1} = \text{constant} \qquad (T7.30)$$

Rich-Context

T7R.1 Suppose that the atmospheric pressure at the top of a tall mountain is about 0.65 times the pressure at sea level. If it is 30°C (86°F) at the beach and a stiff breeze blows this air up the mountain so rapidly that the air essentially expands adiabatically, what is the approximate temperature at the top of the mountain?

T7R.2 In the Los Angeles basin, if the wind blows from the east during the summer, one gets "Santa Ana" winds, which are very hot and dry. This is because air blowing from the high desert to the east of Los Angeles is adiabatically compressed as it descends into the basin. Suppose that it is a nice 90°F summer day in the Mojave desert at an altitude of 4000 ft, but a stiff wind is blowing from the east. What is the air temperature as it reaches Los Angeles (which has an altitude of about 300 ft)? (*Hint:* Assume a model of the atmosphere where the pressure varies with altitude $P \propto e^{-z/z_0}$, where $z_0 \approx 8500$ m: see problem T5R.3.)

T7R.3 Suppose we use a movable piston to confine a sample of nitrogen gas in a cylinder. The cylinder is immersed in ice water, so the gas initially has a temperature of 0°C. (1) You fairly rapidly compress the gas, adiabatically decreasing its volume by a factor of 2. (2) You hold the piston still until the gas has cooled again to 0°C, and then (3) you allow the gas to expand slowly to its original volume while allowing plenty of time for heat to move out of or into the gas (so that its temperature remains very close to 0°C). If 85 g of ice was melted during the second step of the cycle, how much work energy did you put into the gas during the first step? Describe your reasoning. (*Hint:* I suggest that you first draw a PV diagram. Also, it only *looks* as if you do not have enough information to solve this problem. If you find yourself doing lots of calculations, stop and *think* about it some more.)

ANSWERS TO EXERCISES

T7X.1 In this case, you still need to apply an inward external force to the piston to keep the gas confined, but now you let the piston move outward. This means that $\vec{F}_{ext}$ and $d\vec{r}$ are opposite, so the work done in each infinitesimal step is $dW = \vec{F}_{ext} \cdot d\vec{r} = -|\vec{F}_{ext}||d\vec{r}|$. We still have $|\vec{F}_{ext}| \approx |\vec{F}_{gas}| = PA$, but now $dV = A(L_{final} = L_{initial}) = +A|d\vec{r}|$. So $dW = -PA|d\vec{r}| = -P\,dV$, as before.

T7X.2 Substituting as suggested, we get

$$W = -\int_{V_i}^{V_f} \frac{Nk_BT}{V}dV = -Nk_BT\int_{V_i}^{V_f}\frac{dV}{V}$$

$$= -Nk_BT(\ln V_f - \ln V_i) = -Nk_BT\ln\frac{V_f}{V_i} \qquad (T7.31)$$

According to the ideal gas law, $P_iV_i = Nk_BT = P_fV_f$, so we can substitute either P_iV_i or P_fV_f for Nk_BT in equation T7.31.

T7X.3 The chart should look like this:

Step:	Q	W	ΔU
$A \rightarrow B$	+	−	+
$B \rightarrow C$	−	0	−
$C \rightarrow A$	0	+	+

There is more work flowing out of the gas during step $A \rightarrow B$ than into the gas in step $C \rightarrow A$, so the net work is negative. The area inside the cycle in the diagram is about 5.5 squares. Each square corresponds to an energy of

$$\left(20{,}000\,\frac{N}{m^2}\right)(0.05\,m^3)\left(\frac{1\,J}{1\,N\cdot m}\right) = 1000\,J \qquad (T7.32)$$

so $W \approx -5500$ J.

T8

Calculating Entropy Changes

Chapter Overview

Introduction

This chapter develops tools for computing entropy changes in processes involving volume changes and in cases where we do not know an object's multiplicity. These tools will be very useful in chapter T9.

Section T8.1: The Entropy of an Ideal Gas

We can empirically determine the entropy of a gas as follows: (1) We can use the definition of temperature $1/T = dS/dU$ and the empirical result $U = \frac{1}{2}fNk_BT$ to determine how the gas's entropy must depend on energy U. (2) Using the plausible assumption that doubling a molecule's volume doubles the number of microstates available to it, we can determine how the entropy depends on volume V. (3) Noting that the total entropy of two identical gases with number N and volume V is empirically the same as the entropy of a single gas with number $2N$ and volume $2V$ allows us determine how the entropy must depend on N. The final result is

$$\Omega = \left(\frac{U}{N}\right)^{fN/2}\left(\frac{aV}{N}\right)^{N} \quad \Rightarrow \quad S = Nk_B\ln\left[a\frac{V}{N}\left(\frac{U}{N}\right)^{f/2}\right] \tag{T8.7}$$

- **Purpose:** These equations describe the multiplicity Ω and entropy S (respectively) of an ideal gas in a macrostate where its thermal energy is U, its volume is V, and the number of molecules is N, where k_B is Boltzmann's constant, f is the number of a molecule's "degrees of freedom," and a is a constant.
- **Limitations:** The temperature T must be high enough that $U \approx \frac{1}{2}fNk_BT$, the volume per molecule must be much larger than a molecule, and $N \gg 1$. These are typically good assumptions for real gases at everyday temperatures.

Section T8.2: Work and Entropy

The result in the previous section implies that S remains *constant* in an adiabatic expansion or compression. This means that work energy flowing into or out of a gas due to reasonably slow volume changes does not affect the gas's entropy. This is basically because such **quasistatic** volume changes increase or decrease all of the molecules' energy levels proportionally, but do not affect the distribution of molecules in those energy levels, so the multiplicity (the number of possible rearrangements of molecules) among those levels remains the same.

Therefore, if we generalize the equation $dS = dU/T$ (based on the definition of temperature) to read $dS = (dU - W_{qs})/T$ (where W_{qs} is the work done during a quasistatic volume change), then the generalized equation can also handle situations where the system's volume changes. So the part of the energy change dU that *does* affect the entropy is that due to heat dQ, work dW_{other} not related to the quasistatic volume change, and/or other non-heat energy transfers $[dE]$. In the very common case where $[dE] = dW_{other} = 0$, then we simply have

$$dS = \frac{dQ}{T} \qquad \text{(T8.12)}$$

- **Purpose:** This equation tells us how to calculate a system's entropy change dS during an infinitesimal process, where dQ is the infinitesimal heat added to the system, and T is the system's absolute temperature.
- **Limitations:** The heat dQ added must be small enough that T does not change significantly during the process. The system's number of molecules must be constant, its volume must not change or change quasistatically, and it must not experience any energy transfers *other* than heat or the work done by a quasi-static volume change.

Section T8.3: Constant-Temperature Processes

To compute ΔS for a *finite* process (in which the temperature may change), we must integrate equation T8.12: $\Delta S = \int T^{-1} dQ$. This is easy to do for an object if either its temperature or its specific heat is approximately constant during the process. In the first case, we can pull T out in front of the integral to get

$$\Delta S = \frac{Q}{T} \qquad \text{if } T \approx \text{constant during a process} \qquad \text{(T8.14)}$$

This might be the case if (1) the object in question is large enough to be considered a reservoir, (2) it is in good thermal contact with a reservoir, or (3) it is undergoing a **phase change** (then $Q = \pm mL$, where m is the object's mass and L its **latent heat**).

Section T8.4: Handling Changing Temperatures

If an object's specific heat c is approximately constant during the process and its volume does not change significantly, then we can substitute $dQ = mc\,dT$ into the integral (where m is the object's mass), pull mc out of the integral, and integrate with respect to T to get

$$\Delta S = mc \ln \frac{T_f}{T_i} \qquad \text{if no work is done and } c \approx \text{constant} \qquad \text{(T8.25)}$$

Section T8.5: Replacement Processes

So we see that we can easily calculate entropy changes for three basic processes:

1. Quasistatic adiabatic processes ($\Delta S = 0$).
2. Quasistatic isothermal processes ($\Delta S = Q/T$).
3. Constant-volume heating where $c \approx$ constant [$\Delta S = mc \ln(T_f/T_i)$].

But a system's entropy by definition depends only on its macrostate. Therefore, the *change* in a system's entropy depends only on the system's initial and final macrostates, and not at all on the process that gets it from one to the other. This means we can calculate the entropy change involved in *any* process (even if the process does not fit into one of the categories above) if we can find a **replacement** process involving a sequence of processes that *do* all fit into the categories above and that take us from the same initial macrostate to the same final macrostate. The entropy change we compute for the replacement process will be the same as that for the original process.

T8.1   The Entropy of an Ideal Gas

Our goal: determine a gas's
entropy empirically

Chapter T6 provides us with the tools we need to derive an expression for
the multiplicity Ω (and thus the entropy $S = k_B \ln \Omega$) of an ideal gas, and problem T8D.1 works through the derivation if you are interested. But we can-
not describe most realistic macroscopic objects using models that are simple
enough to allow us to actually *calculate* entropies. Rather, we must experimentally *measure* their entropies. This chapter is all about how one can determine entropy changes by using macroscopic measurements. We will start
this section by discussing how we might use macroscopic reasoning to infer
the entropy of an ideal gas, which we can compare to the theoretical result.

We define entropy $S = k_B \ln \Omega$, where Ω is the number of microstates consistent with a system's macrostate. We saw in chapter T7 that three macroscopic variables suffice to specify a gas's macrostate: we will choose those
variables here to be the gas's thermal energy U, its volume V, and the number
of molecules N. Now, the starting point for measuring *any* system's entropy
is to invert the definition of temperature $1/T = \partial S / \partial U$ presented in chapter
T3. If we know the (infinitesimal) amount of energy dU we have added to
the system at a known absolute temperature T, then we can calculate the
system's (infinitesimal) change in entropy dS as follows:

$$dS = \frac{dU}{T} \tag{T8.1}$$

But the definition of temperature (and thus equation T8.1) applies *only* when
V and N remain constant, so we must keep V and N constant during the process where we add energy dU. But this means that equation T8.1 is useless for
determining how a system's entropy depends on V and N.

How S must depend on U

But we can (at least!) determine how S depends on U. Suppose we know
the entropy S_0 at some reference macrostate with reference energy U_0 and
want to know how S varies with U when we hold V and N fixed. We know
that for an ideal gas, $dU = \frac{1}{2} f N k_B dT$, where f is the gas's number of "degrees
of freedom." Substituting this into equation T8.1 and integrating yields

$$S - S_0 = \int_{T_0}^{T} \frac{\frac{1}{2} f N k_B \, dT}{T} = \frac{f}{2} N k_B \int_{T_0}^{T} \frac{dT}{T} = \frac{f}{2} N k_B [\ln T - \ln T_0] \tag{T8.2a}$$

where T_0 is the gas's temperature in the initial macrostate. But note that for an
ideal gas, $U = \frac{1}{2} f N k_B T \Rightarrow T = 2U/f N k_B$. Moreover, $\ln x - \ln y = \ln(x/y)$, so

$$S = S_0 + \frac{f}{2} N k_B \ln \left(\frac{T}{T_0} \right) = S_0 + \frac{f}{2} N k_B \ln \left(\frac{2U/f N k_B}{2U_0/f N k_B} \right) = \frac{f}{2} N k_B \ln \left(\frac{U}{N} \right) + A \tag{T8.2b}$$

where $A \equiv S_0 - \frac{1}{2} f N k_B \ln (U_0/N)$ is a quantity that does not depend on U. But
since $S \equiv k_B \ln \Omega$, the gas's multiplicity Ω must be

$$\Omega(U,V,N) = e^{S/k_B} = e^{A/k_B} \left(\frac{U}{N} \right)^{fN/2} = B(V, N) \left(\frac{U}{N} \right)^{fN/2} \tag{T8.3}$$

where $B \equiv e^{A/k_B}$ is a factor that may depend on V or N, but not on U.

How S must depend on V

Now let's turn to the question of how the gas's multiplicity depends
on V. Suppose that we use a barrier to confine a gas to half of a container
(see figure T8.1a). If we remove the barrier, then the gas will spontaneously
expand into the other part of the container. (Because the gas does no work
when expanding into a vacuum, the gas's energy U remains constant in this
process.) This process is irreversible, so increasing the volume available to
the gas *must* increase its multiplicity and thus its entropy even though both
its energy U and the number of gas molecules N remain constant.

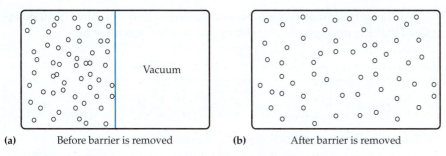

(a) Before barrier is removed **(b)** After barrier is removed

Figure T8.1
Free expansion of a gas into a vacuum.

Indeed, we might guess intuitively that doubling the volume available to a given molecule should double the number of microstates available to it (other things being equal). If *each* molecule's microstates are doubled by this expansion, then the number of microstates available to the whole system must go up by a factor of 2^N. More generally, we should have

$$\Omega \propto \left(\frac{V}{V_0}\right)^N \quad \Rightarrow \quad \Omega = F(N)\left(\frac{U}{N}\right)^{fN/2} V^N \tag{T8.4}$$

where V_0 is some reference volume and $F(N)$ is some function of N alone.

Now consider the situation shown in figure T8.2. Here we have a barrier that divides the container in half, but both halves contain identical gases. When we remove or replace the barrier, the situation remains in equilibrium, so the total multiplicity must remain unchanged. Let U, N, and V be the thermal energy, number of molecules, and volume of the gas in each half. Because the combined system's multiplicity is the product of the multiplicities of each part, and because that multiplicity must remain the same whether the barrier is present or not, we must have

How S must depend on N

$$\Omega_{tot} = F(2N)\left(\frac{2U}{2N}\right)^{fN}(2V)^{2N} = \left[F(N)\left(\frac{U}{N}\right)^{fN/2} V^N\right]\left[F(N)\left(\frac{U}{N}\right)^{fN/2} V^N\right]$$

$$\Rightarrow \quad F(2N)2^{2N} = [F(N)]^2 \tag{T8.5}$$

You can ask WolframAlpha to solve this tricky little equation for the function $F(N)$, or you can easily check that $F(N) \equiv (a/N)^N$ (where a is an arbitrary constant) does the trick:

$$\Omega_{tot} = \left(\frac{2U}{2N}\right)^{fN}\left(\frac{a2V}{2N}\right)^{2N} = \left[\left(\frac{U}{N}\right)^{fN/2}\left(\frac{aV}{N}\right)^N\right]\left[\left(\frac{U}{N}\right)^{fN/2}\left(\frac{aV}{N}\right)^N\right] \tag{T8.6}$$

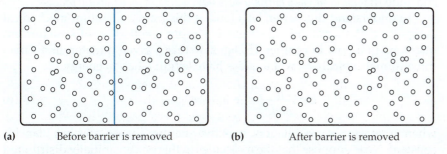

(a) Before barrier is removed **(b)** After barrier is removed

Figure T8.2
Removing a barrier that separates a gas into two identical halves.

So, using basic macroscopic observations and reasoning, we have determined that the multiplicity and entropy of an ideal gas must be

The entropy of an ideal gas

$$\Omega = a^N \left(\frac{U}{N}\right)^{fN/2} \left(\frac{V}{N}\right)^N \quad \Rightarrow \quad S = Nk_B \ln\left[a\frac{V}{N}\left(\frac{U}{N}\right)^{f/2}\right] \tag{T8.7}$$

- **Purpose:** These equations describe the multiplicity Ω and entropy S (respectively) of an ideal gas in a macrostate where its thermal energy is U, its volume is V, and the number of molecules is N, where k_B is Boltzmann's constant, f is the number of a molecule's energy-storage "degrees of freedom," and a is an unknown constant.
- **Limitations:** The temperature T must be high enough that $U \approx \frac{1}{2}fNk_BT$, V/N must be much larger than a molecule, and $N \gg 1$. These are typically good assumptions for real gases at everyday temperatures.

Indeed, a basic quantum model of the ideal ideal monatomic gas for large U and large N implies (see problem T8D.1) that the gas's entropy is given by the so-called **Sackur-Tetrode equation:**

The Sackur-Tetrode equation

$$S = Nk_B \ln\left[\frac{V}{N}\left(\frac{4\pi e^{5/3}\, m}{3h^2}\frac{U}{N}\right)^{3/2}\right] \tag{T8.8}$$

where m is the mass of a molecule, h is Planck's constant, and $e = 2.718\ldots$ (not the charge of the proton). We see that the only new thing that the quantum model provides is the value $a = (4\pi e^{5/3}m/3h^2)^{3/2}$ of the unknown constant in equation T8.7. But this constant rarely matters for practical applications.

T8.2   Work and Entropy

Entropy does not change in an adiabatic process

Consider now a fixed amount of gas (N = constant) undergoing a process where its volume V changes but its entropy S remains fixed. Because increasing a gas's volume will increase the number of microstates available to each molecule, the gas's thermal energy U must decrease to compensate. Indeed, equation T8.7 implies that if S is constant in such a process, we must have

$$\text{constant} = VU^{f/2} = V(\tfrac{1}{2}fNk_BT)^{f/2} \quad \Rightarrow \quad \text{constant}' = VT^{f/2}$$

$$\Rightarrow \quad \text{constant}'' = V^{2/f}T = V^{\gamma-1}T \tag{T8.9}$$

because the adiabatic index $\gamma \equiv 1 + 2/f$. But we saw in chapter T7 that during an adiabatic process $TV^{\gamma-1}$ is constant (see equation T7.16). We can therefore conclude that *an ideal gas's entropy does not change in an adiabatic process.*

This is because such a process changes the *states'* energies without changing the possibilities for rearranging particles

To understand this better, let's look at this from a quantum-mechanical perspective. A monatomic ideal gas is simply a collection of structureless quantons in a box. In chapter T5, we saw that a given quanton in a three-dimensional cubical box whose sides have length L has an energy

$$E = \frac{h^2}{8mL^2}(n_x^2 + n_y^2 + n_z^2) \tag{T8.10}$$

where n_x, n_y, and n_z are integers, m is the quanton's mass, and h is Planck's constant. Now, suppose that the molecules in this box are initially distributed in a certain way among the various energy levels and that the gas's multiplicity under these circumstances is Ω_0, reflecting the numbers of ways that we

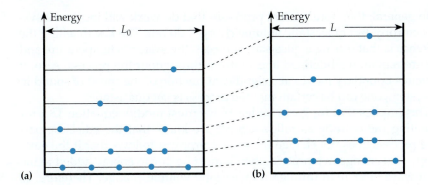

Figure T8.3
In these diagrams, the black outline represents a box holding gas molecules, the thin horizontal lines various quantum energy states, and the colored dots some molecules distributed over those states. Compressing the box proportionally increases the energies of the states (and thus the energies of the molecules in those states) without affecting the number of ways we could rearrange the molecules to get the same (proportionally larger) total energy.

might rearrange the molecules among the energy states without changing the gas's total energy U_0. Now suppose that we slowly squeeze the box, decreasing its volume V. Pretend for the sake of argument that during this process, the gas's molecules remain in exactly the same quantum states (each molecule's values of n_x, n_y, and n_z remain the same). However, because $L^2 = V^{2/3}$ appears in the denominator of equation T8.10, squeezing the box proportionally increases the energies of *all* of the molecules' quantum states (see figure T8.3). Even though the system's total energy increases to U (equal to the amount of work we do to compress the gas), the gas's multiplicity must remain the *same* in this process, because we still have *exactly* the same number of ways that we could rearrange the molecules among the (proportionally higher) energy states to get the (proportionally higher) total energy. So, the work energy that flows into the gas during this compression therefore goes not to bumping the molecules to higher (but fixed) energy levels (which *would* increase the multiplicity), but rather to increasing the height of the energy levels themselves (which *doesn't* increase the multiplicity).

One can check this hypothesis experimentally. If we compress a gas adiabatically, and then allow it to expand adiabatically back to its original volume, we find that (if we do this sufficiently slowly) that the gas spontaneously returns to its original thermal energy and temperature. The fact that the process is experimentally reversible offers compelling physical evidence that the gas's entropy has remained constant. Because (in principle) we could perform the compression and expansion with, say, a piston connected to a frictionless flywheel, whose entropy does not change in the purely mechanical process of rotation, the fact that this adiabatic volume change is reversible means that the gas's entropy doesn't change either.

This is an example of a more general truth: when a system absorbs work energy from its surroundings or does work on those surroundings in a sufficiently slow process, the work energy flow does not *in itself* change either the system's entropy or that of its surroundings. The process must, however, be sufficiently slow (the technical term is **quasistatic**) for this to be true. To see why, imagine that I compress a gas very suddenly (so suddenly that the piston's velocity is a significant fraction of the speed of sound). In such a case, molecules will pile up against the rapidly moving piston, causing the pressure against the piston to be larger than that elsewhere in the gas. We would therefore have to do more work on the gas during this sudden compression than we would if we compressed it slowly. As this pressure wave (moving at the speed of sound) redistributes this extra energy to the rest of the gas, it must bump molecules up into higher energy states, since the increase in the energy levels themselves can no longer absorb all the work energy. Actually bumping molecules to higher energy levels *will* increase the multiplicity and thus the entropy of the gas.

General truth: quasistatic work can change a system's energy without changing its entropy

In general, therefore, violent processes that do work will increase a system's entropy, but quasistatic volume changes do not. The key to telling the difference is that during a quasistatic process, the system's temperature and pressure remain well-defined and uniform throughout the process. Also, if a system's boundary moves at a significant fraction of the speed of sound in that system, one can be certain that the process is *not* quasistatic.

This implies that for gases (at least), we must modify equation T8.1 for calculating the entropy as follows: Since we know that the adiabatic part of the process does not change the gas's entropy, it follows that if the gas's entropy *does* change in a process, then it must be due to energy added to the gas in ways *other* than through a quasistatic volume change. Therefore, as long as N is constant and V changes only quasistatically, then

$$dS = \frac{dU - dW_{qs}}{T} \tag{T8.11}$$

where dW_{qs} is the work associated with the quasistatic volume change. In many practical gas processes, dW_{qs} is the *only* kind of energy added to the gas other than heat dQ, so in such cases,

A basic formula to use to compute entropy changes

$$dS = \frac{dQ}{T} \tag{T8.12}$$

• **Purpose:** This equation tells us how to calculate a system's entropy change dS during an infinitesimal process, where dQ is the infinitesimal heat added to the system, and T is the system's temperature.
• **Limitations:** The heat dQ added must be small enough that T does not change significantly during the process. The system's number of molecules must be constant, its volume must either not change or change quasistatically, and it must not experience any energy transfers *other* than heat or the work done by a quasistatic volume change.

We will spend the rest of the chapter exploring the uses of equation T8.12.

Though we derived equation T8.12 in the context of gases, it works for solids and liquids as well. Since most solids and liquids change volume only slightly, if at all, during typical processes, we can treat V as fixed. If N is also fixed, then equation T8.11 applies. However, as long as no *other* kind of work energy enters the solid or liquid in question, then $dU = dQ$, meaning that equations T8.11 and T8.12 are equivalent.

Please note the limitations on this equation

You should carefully note equation T8.12's many limitations. Surprisingly, we will find that with a bit of cleverness, we can sidestep these limitations and use equation T8.12 to calculate dS even for violent processes or processes that involve other kinds of energy transfers. We'll see how in section T8.5.

Exercise T8X.1

Which of the following processes are consistent with the limitations on equation T8.12, and which are not?

(a) A stone is thrown into a pond.
(b) Soup is slowly heated on a stove.
(c) A cup of coffee is gently stirred.
(d) A bubble slowly rises in a lake.
(e) A gas is slowly compressed in a cylinder while its temperature is held fixed.

T8.3   Constant-Temperature Processes

Equation T8.12 strictly applies only to heat transfers dQ that are "sufficiently small" that the temperature does not change significantly during the process. In general, as we transfer heat to an object, its temperature will change. We cannot calculate ΔS for a finite heat transfer ΔQ unless we take account of the changing temperature by doing an integral:

$$dS = \frac{dQ}{T} \quad \Rightarrow \quad \Delta S = \int_{\text{process}} \frac{dQ}{T} \qquad \begin{array}{l}\text{quasistatic volume}\\ \text{change, no other}\\ \text{work done, } N \text{ fixed}\end{array} \qquad \text{(T8.13)}$$

We will discuss how to evaluate such integrals in section T8.4.

On the other hand, if the object's temperature remains nearly *constant* during the process, then whatever heat is transferred during the process is automatically "sufficiently small" and we can apply equation T8.12 directly:

$$\Delta S = \frac{Q}{T} \qquad \text{if } T \approx \text{constant during a process} \qquad \text{(T8.14)}$$

The entropy change when temperature is approximately constant

In what circumstances will T be approximately constant? In chapter T1, we discussed how the change dU in an object's thermal energy as its temperature changes by dT is given by

$$dU = mc\, dT \qquad \text{(T8.15)}$$

where m is the object's mass and c is its specific heat (note that specific heats for various substances are listed on the inside front cover of this text). Consider an object so massive that it can absorb or supply a significant amount of heat dU while suffering only the tiniest change in temperature dT. The technical term for such an object in thermal physics (as we saw in chapter T4) is a **reservoir**. Thus, we can use T8.14 to

1. Compute the entropy change of a *reservoir* absorbing or supplying heat.
2. Compute the entropy change of something in thermal contact with a reservoir, and thus whose temperature is the same as that of the reservoir.

Three practical situations where $T \approx$ constant

In both cases, the reservoir ensures that the temperature T appearing in equation T8.14 does not change significantly during the process.

There is a third (unrelated) case in which we can use equation T8.14. We saw in unit C that the internal latent energy we must add to or remove from a substance while it undergoes a **phase change** (for example, from a solid to a liquid) is given by

$$\Delta U^{\text{la}} = \begin{cases} -mL \\ +mL \end{cases} \quad \begin{array}{l}\text{(gas to liquid, liquid to solid)}\\ \text{(solid to liquid, liquid to gas)}\end{array} \qquad \text{(T8.16)}$$

where m is the object's mass and L is the **latent heat** for the substance and phase change (latent heats for various substances are also listed on the inside front cover). Since the substance's temperature remains *constant* until the phase change is complete, then if we supply or remove this energy from the system in the form of heat, we can use equation T8.14 to

3. Compute the entropy change of a substance during a phase change.

These three cases therefore comprise the most common practical situations in which we can apply equation T8.14. The following examples illustrate its use.

Example T8.1

Problem: Suppose we have a bathtub full of water at 20°C and we place a 1.0-kg stone in it whose original temperature is 95°C. What is the entropy change of the water in this case?

Solution The tub of water probably contains hundreds of kilograms of water, so its temperature change in this process will be tiny compared to that of the stone. This means that the final equilibrium temperature of both will be about 20°C. According to the inside front cover, the specific heat of granite is about 760 J kg^{-1} K^{-1}. Let Q_w and Q_s be the heat flowing into the water and stone, respectively, and let T_w and T_s be the temperatures of the water and stone, respectively. The heat energy Q_s flowing into the stone is

$$\Delta Q_S = m_s c_s \Delta T_s = (1.0 \text{ kg})(760 \text{ J kg}^{-1} \text{K}^{-1})(-75 \text{ K}) = -57{,}000 \text{ J} \qquad \text{(T8.17)}$$

This is negative, so heat is actually flowing *out* of the stone. The water gains this energy, so the water's change in entropy is

$$\Delta S = \frac{\Delta Q_w}{T_w} = \frac{+|\Delta Q_s|}{T_w} = \frac{+57{,}000 \text{ J}}{293 \text{ K}} = +195 \text{ J/K} \qquad \text{(T8.18)}$$

This is positive, as we would expect for something gaining thermal energy.

Example T8.2

Problem: Suppose we have a cylinder containing an ideal gas in good thermal contact with a reservoir at 32°C. Imagine that we slowly compress the gas, doing 45 J of work on it while its temperature remains constant. What is the change in the gas's entropy? In that of the reservoir?

Translation Let Q_g and Q_R be the heat flowing into the gas and reservoir, respectively, and let T_g and T_R be the temperatures of the gas and reservoir, respectively. Because the gas's temperature does not change, its thermal energy U does not change. Therefore, any work energy that it gains in this compression must flow out to the reservoir in the form of heat. The gas therefore *loses* 45 J of heat in this process, so its entropy change [noting that $T_g = (273 + 32)$ K = 305 K] is

$$\Delta S_g = \frac{Q_g}{T_g} = \frac{-45 \text{ J}}{305 \text{ K}} = -0.15 \text{ J/K} \qquad \text{(T8.19)}$$

(Note that the work energy it gains in the "slow" compression doesn't count!) The 45 J that the gas loses flows *into* the reservoir, so its entropy change is

$$\Delta S_R = \frac{Q_R}{T_R} = \frac{+45 \text{ J}}{305 \text{ K}} = +0.15 \text{ J/K} \qquad \text{(T8.20)}$$

Note that the *net* entropy change of the system in this process is zero.

Example T8.3

Problem: Suppose 120 g of ice at 0°C melts to a puddle of water at 0°C on a surface in a room where the temperature is 28°C. What is the change in the water's entropy?

Solution Let Q_w and T_w be the heat flowing into the ice and its temperature, respectively. Let m be the mass of the ice and L its latent heat. Note that the temperature of the ice does not change as it melts to water, so we can use equation T8.14. According to the table on the inside front cover, the latent heat associated with the transformation of solid to liquid water is 333 kJ/kg, so the total heat that the ice must absorb from the warm room is

$$Q_w = dU = +mL = (0.12 \text{ kg})(333 \text{ kJ/kg}) = 40 \text{ kJ} \tag{T8.21}$$

Its entropy change is therefore

$$\Delta S_w = \frac{Q_w}{T_w} = \frac{+40{,}000 \text{ J}}{273 \text{ K}} = +150 \text{ J/K} \tag{T8.22}$$

(The room's temperature is irrelevant here.) Note that the entropy change is positive, as we should expect when a system's internal energy increases.

Note that in examples T8.1 through T8.3 I carefully converted temperatures to *kelvins* before computing the entropy. If I had divided by temperatures in degrees Celsius, I would have gotten *very* different answers. We *must* use absolute temperatures when we use equation T8.14 because that equation is ultimately based on the formula for the definition $1/T = \partial S/\partial U$ of the *absolute* temperature scale.

T8.4   Handling Changing Temperatures

Now suppose we need to compute an object's entropy change during a process that *changes* the object's temperature significantly. Assume the object has specific heat c and mass m. If the object's volume does not change (much) during the process, then no significant work is done and

$$dQ \approx dU = mc \, dT \qquad \text{if no work is done} \tag{T8.23}$$

Substituting this into equation T8.13 yields

$$\Delta S = \int_{\text{process}} \frac{dQ}{T} = \int_{\text{process}} \frac{mc \, dT}{T} \qquad \text{if no work is done} \tag{T8.24}$$

Note that if we consider c to be a function of temperature, this is simply an integral over some function of the object's changing temperature T.

    If the object's specific heat is approximately independent of temperature (as it often is over reasonably small temperature ranges), then we can pull both m and c out in front of the integral. Integrating the dT/T that remains from the object's initial temperature T_i to its final temperature T_f yields

$$\Delta S = mc \int_{T_i}^{T_f} \frac{dT}{T} = mc \left[\ln T \right]_{T_i}^{T_f} = mc[\ln T_f - \ln T_i]$$

$$= mc \ln\!\left(\frac{T_f}{T_i}\right) \qquad \text{if no work is done and } c \approx \text{constant} \tag{T8.25}$$

The change in entropy for a heat exchange with no work involved

where in the last step I have used the fact that $\ln x - \ln y = \ln(x/y)$.

    This expression is very useful for solids and liquids whose volume does not change significantly as their temperature changes. It is also useful for gases in isochoric (constant-volume) processes. But since equation T8.23 does not apply to gases undergoing any kind of volume change, we cannot use equation T8.25 either.

    The following example illustrates an application of this equation.

Example T8.4

Problem: Suppose that we place an aluminum block with mass $m_b = 260$ g and an initial temperature $T_b = 89°C$ into a cup of water with mass $m_w = 320$ g and an initial temperature $T_w = 22°C$, and I allow the two to come to thermal equilibrium. What is the entropy change of the water in this process?

Solution Let $c_b = 900$ J/(kg·K) and $c_w = 4186$ J/(kg·K) (see the inside front cover) be the specific heats of the aluminum block and water, respectively. If we assume that these values are reasonably constant over the temperature ranges involved, and we assume that the volume changes of the aluminum and water are negligible, then we can use equation T8.25. The first step is to find the final equilibrium temperature T_f. The heat flowing out of the metal goes into the water; so if Q_b and Q_w represent the heat flowing into the block and water, respectively, then $Q_b = -Q_w$. This means that

$$m_b c_b(T_f - T_b) = Q_b = -Q_w = -m_w c_w(T_f - T_w) \qquad (T8.26)$$

If we solve for T_f, we get

$$(m_w c_w + m_b c_b)T_f = m_w c_w T_w + m_b c_b T_b \quad \Rightarrow \quad T_f = \frac{m_w c_w T_w + m_b c_b T_b}{m_w c_w + m_b c_b} \qquad (T8.27)$$

If we convert $T_w = 22°C$ to $(273 + 22)$ K $= 295$ K (similarly, $T_b = 362$ K), we get

$$T_f = \frac{(0.32\ \text{kg})(4186\ \text{J·kg}^{-1}\text{K}^{-1})(295\ \text{K}) + (0.26\ \text{kg})(900\ \text{J·kg}^{-1}\text{K}^{-1})(362\ \text{K})}{(0.32\ \text{kg})(4186\ \text{J·kg}^{-1}\text{K}^{-1}) + (0.26\ \text{kg})(900\ \text{J·kg}^{-1}\text{K}^{-1})}$$

$$= 305\ \text{K} = 32°C \qquad (T8.28)$$

(See problem T1D.2 for another way to compute the equilibrium temperature.) Since T_f is significantly different from the water's initial temperature of 22°C, we need equation T8.25 to compute the water's entropy change

$$\Delta S_w = m_w c_w \ln\left(\frac{T_f}{T_w}\right) = (0.32\ \text{kg})(4186\ \text{J·kg}^{-1}\text{K}^{-1})\ln\left(\frac{305\ \text{K}}{295\ \text{K}}\right) = +45\ \frac{\text{J}}{\text{K}} \quad (T8.29)$$

The units are correct, and the water's entropy change is positive. This is plausible, because the water is gaining energy here.

Exercise T8X.2

Did we *have* to use absolute temperatures in equation T8.28 (that is, would using temperatures in degrees Celsius make any difference in the final result)? Did we have to use absolute temperatures in equation T8.29?

Exercise T8X.3

Show that the block's entropy change is −40 J/K. (This means that the net entropy change for the block/water system is +5 J/K: An *increase* is to be expected in a spontaneous heat transfer process, because such a process is irreversible.)

T8.5   Replacement Processes

Equations T8.12 and its descendants T8.14 and T8.25 do *not* apply to processes in which volume changes are not quasistatic or if work other than that resulting from quasistatic volume changes is done during the process. In many cases of such processes, however, we *can* actually use these equations, employing a clever argument to do a kind of "end run" around the difficulties presented by the real process.

Our clever argument hinges on the fact that the entropy of any system is defined in terms of the current *macrostate* of that system. Remember that $S \equiv k_B \ln \Omega$ where Ω is the number of microstates that are consistent with the system's macrostate. This number depends only on the characteristics of the system's macrostate, not how the system got to that macrostate. Therefore, any change in entropy S that the system experiences as it goes from one macrostate to another *will depend on that system's initial and final macrostates alone, and not at all on the process by which we got from one to the other!*

So consider a system that goes from one macrostate to another via a process that is *not* quasistatic and/or violates the conditions of the equations that we have at hand. If we can imagine *any* process satisfying the restrictions on equation T8.12 that takes us from the same initial macrostate to the same final macrostate, we can calculate the entropy change using this hypothetical process (which we will call a **replacement process**) instead of the actual process: we *must* get the same answer that we would have gotten if we could have done the calculation for the actual process (see figure T8.4).

This is a *very* powerful technique that allows us to handle virtually any situation we can imagine. We can almost always replace even the most violent or crazily complex processes (as long as the process begins and ends in a well-defined macrostate) with some sequence of simple and quasistatic processes for which we can easily calculate the change in entropy.

The following examples illustrate some different ways we can apply this very useful principle.

An object's entropy depends on its macrostate alone

Using a *replacement process* to calculate an entropy change

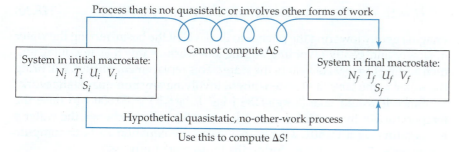

Process that is not quasistatic or involves other forms of work

Cannot compute ΔS

System in initial macrostate:
$N_i \ \ T_i \ \ U_i \ \ V_i$
S_i

System in final macrostate:
$N_f \ \ T_f \ \ U_f \ \ V_f$
S_f

Hypothetical quasistatic, no-other-work process

Use this to compute ΔS!

Figure T8.4

The change of entropy ΔS must be the same for *any* process that goes from a given initial to a given final macrostate. So we can calculate the entropy change ΔS for a process that violates the restrictions on equation T8.12 by replacing the process by a different process that satisfies those restrictions and still takes us between the desired initial and final states.

Example T8.5

Problem: Suppose we pour a pitcher containing 2 kg of water at 22°C from a height of 0.3 m into a basin containing 3 kg of water at the same temperature (see the picture below). What is the entropy change of the 5 kg of water?

(Credit: © McGraw-Hill Education/Mark Dierker, photographer)

Solution When the 2 kg of water falls into the basin, its gravitational potential energy is converted to kinetic energy and then to thermal energy in the basin. This thermal energy gain is *not* due to a quasistatic volume change, nor is it heat entering the system (there are no temperature differences involved). Therefore, equation T8.12 does not apply.

The initial macrostate of our system is 5 kg of water (3 kg in the basin and 2 kg in the pitcher) at $T = 22$°C and thermal energy U_i. Our final state is the same mass of water having a slightly higher thermal energy of

$$U_f = U_i + mgh = U_i + (2 \text{ kg})(9.8 \text{ m/s}^2)(0.3 \text{ m}) = U_i + 5.9 \text{ J} \qquad \text{(T8.30)}$$

Imagine gently lowering the pitcher to the level of the basin, mixing the water in slowly and gently (avoiding adding *any* work as much as possible), and then adding $Q = 5.9$ J of heat to the water. This replacement process will bring the water to the *same* final state without involving any non-quasistatic work.

Since it would take $(5 \text{ kg})(4186 \text{ J·kg}^{-1}\text{K}^{-1})(1 \text{ K}) > 20{,}000$ J to raise the temperature of the water by even 1 K, our 5.9 J will not change the water's temperature significantly. Therefore, we can use equation T8.14 to compute the water's entropy change during the replacement process:

$$\Delta S = \frac{Q}{T} = \frac{+5.9 \text{ J}}{(273 + 22)\text{K}} = +0.020 \text{ J/K} \qquad \text{(T8.31)}$$

The water's entropy change during the actual process *must* be the same.

Problem: Use a replacement process to calculate the entropy change for the sudden expansion process shown in figure T8.1, where removing a barrier allows the gas to expand in a vacuum to twice its original volume.

Solution The volume change here is not even remotely quasistatic, so equation T8.12 does not directly apply. However, since the gas expands into a vacuum, it does no work and there is no time for heat to flow, so the gas's thermal energy U is unchanged in this process. The initial and final macrostates of the gas are thus specified by U, V, N and U, $2V$, N, respectively, where N is the fixed number of molecules involved. One replacement process that takes us from the same initial macrostate to the same final macrostate is the following. Suppose that we use a piston to gradually allow the gas to expand to its final volume. As the gas expands against the piston, though, work energy will flow out of the gas, leading to a decrease in its thermal energy U. We want the final energy of the gas to be the same as it was originally, so we need to add heat to replace the work energy lost. We can do this in a particularly easy way by putting the gas in thermal contact with a reservoir at 22°C. Since the thermal energy U of an ideal gas depends on T and N but not V, keeping the gas's temperature fixed will automatically add whatever heat is needed to keep its internal energy U fixed. If we calculate the entropy change for this replacement process, it should be the same as that for the original process.

The heat Q that we must add in this replacement process must be equal to the work that the gas does as it expands isothermally. Equation T7.10 gives the work for an isothermal process, so according to that equation, we have

$$Q = -W = +Nk_B T \ln\left(\frac{2V}{V}\right) = Nk_B T \ln 2 \qquad (T8.32)$$

Since the temperature of the gas is a constant 22°C = 295 K in our replacement process, we can now use equation T8.14 to calculate the change in the gas's entropy:

$$\Delta S = \frac{Q}{T} = +\frac{Nk_B T}{T} \ln 2 = Nk_B \ln 2 \qquad (T8.33)$$

This is completely consistent with what we would find by using equation T8.7 to calculate the entropy change directly:

$$\Delta S = Nk_B \ln\left[a\frac{2V}{N}\left(\frac{U}{N}\right)^{f/2}\right] - Nk_B \ln\left[a\frac{V}{N}\left(\frac{U}{N}\right)^{f/2}\right]$$

$$= Nk_B\left[\ln a + \ln 2 + \ln\left(\frac{V}{N}\right) + \frac{f}{2}\ln\left(\frac{U}{N}\right)\right] - Nk_B\left[\ln a + \ln\left(\frac{V}{N}\right) + \frac{f}{2}\ln\left(\frac{U}{N}\right)\right]$$

$$= Nk_B \ln 2 \qquad (T8.34)$$

where I have used $\ln xy = \ln x + \ln y$. (Note that most of the terms cancel out, because we are only changing the volume.) This is a nice check that our replacement process really does give us the right entropy change!

TWO-MINUTE PROBLEMS

T8T.1 A container holds 1 million helium molecules at 300 K. If we double the gas's temperature to 600 K, by about what factor does its multiplicity increase?
A. $\ln 2$
B. 2
C. $2^{3/2}$
D. $10^6 k_B \ln 2$
E. $2^{1,000,000}$
F. $2^{1,500,000}$
T. Some other factor (specify)

T8T.2 How does the entropy of 1 mol of helium gas in a container at room temperature compare to 1 mol of argon gas in an identical container at the same temperature?
A. The helium gas has greater entropy.
B. The argon gas has greater entropy.
C. Both gases have the same entropy.
D. One needs to know more to answer definitively.

T8T.3 Suppose that we allow an ideal gas containing N molecules to expand slowly and adiabatically to twice its initial volume. By what factor does its *multiplicity* increase?
A. By a factor of 2.
B. By a factor of 2^N.
C. The multiplicity does not change.
D. The answer depends on the gas's initial volume.
E. The answer depends on the gas's initial energy.

T8T.4 Suppose that we allow an ideal gas containing N atoms to expand slowly and isothermally to twice its initial volume. By what factor does its *multiplicity* increase?
A. By a factor of 2.
B. By a factor of 2^N.
C. The multiplicity does not change.
D. The answer depends on the gas's initial volume.
E. The answer depends on the gas's fixed temperature.

T8T.5 Suppose we hold a sample of helium gas at a constant temperature while we slowly compress it in a cylinder until its volume has decreased by a factor of 4. During this process, the gas's entropy
A. increases.
B. decreases.
C. remains the same.
D. Does something else (specify).

T8T.6 Consider each of the following processes. Is the process consistent (C) or inconsistent (D) with the limitations on equation T8.12 (that N be fixed and that the process not involve any energy transfer that is neither heat nor work due to a quasistatic volume change)?
a. A glass of milk is spilled on the floor.
b. A closed vial of gas is heated with a flame.
c. A pan of water boils on the stove.
d. Sunlight is absorbed by a black hat.
e. Water slowly evaporates from your skin.

T8T.7 Suppose that a gas expands in a cylinder, pushing back a piston with a speed of 20 m/s (about 45 mi/h). We can reasonably consider such an expansion to be quasistatic. T or F?

T8T.8 In a tub, 100 kg of water at 30°C absorbs 100 J from an electric heater. We can reasonably consider the water to be a reservoir in this process. T or F?

T8T.9 We slowly heat 100 g of water from 5 to 25°C. What is its change in entropy? (For water, $c = 4186$ kg.)
A. 582 J/K
B. 29.2 K
C. 139 J/K
D. 6.95 J/K
E. 0
F. Other (specify)

T8T.10 We bring a hot object into contact with a cold object. The hot object's entropy decreases by 1.5 J/K as it comes into equilibrium with the cold object. By how much does the cold object's entropy increase?
A. More than 1.5 J/K
B. Exactly 1.5 J/K
C. Less than 1.5 J/K
D. 0

T8T.11 We drop a stone of mass m from a height h into a bucket of water. What would be a suitable replacement process for this non-quasistatic process?
A. Lower the stone gently into the water.
B. Lower the stone gently into the water, and then heat the water until it has mgh more energy than it had before.
C. Lower the stone gently into the water, and then stir the water gently to give it mgh more energy.
D. Raise the water up to height h, gently put the stone in it, and then lower both back to the ground.
E. Either (B) or (C).
F. None of the above.

T8T.12 A sample of helium gas with an initial temperature of T_i in an insulated cylinder with initial volume V_i is suddenly and violently compressed to one-half its volume. The absolute temperature of the gas after the compression is observed to have doubled. (Note: In a *quasistatic* adiabatic compression of a monatomic gas, $TV^{2/3} = $ constant.) A suitable replacement process for this process would be
A. a slow isothermal compression from V_i to $\frac{1}{2}V_i$.
B. a slow adiabatic compression from V_i to $\frac{1}{2}V_i$.
C. Heating the gas to double its temperature followed by a slow adiabatic compression from V_i to $\frac{1}{2}V_i$.
D. a slow adiabatic compression from V_i to $\frac{1}{2}V_i$ followed by heating to bring the temperature to $2T_i$.
E. a slow adiabatic compression from V_i to $\frac{1}{2}V_i$ followed by cooling to bring the temperature to $2T_i$.
F. None of the above.

HOMEWORK PROBLEMS

Basic Skills

T8B.1   Starting with equation T8.8, find an expression for the entropy of an ideal monatomic gas as a function of T, V, and N instead of U, V, and N.

T8B.2   Suppose we compress a sample of N molecules of helium gas to one-half its original volume and simultaneously raise its absolute temperature by a factor of 2. Use equation T8.8 to calculate the change in the gas's entropy in this process as a multiple of Nk_B. (*Hint: $U \propto T$.*)

T8B.3   Suppose we add 12 J of heat to 10 kg of water at 15°C in a tub.
(a) Argue that this amount of energy will not change the water's temperature very much.
(b) Compute the water's entropy change.

T8B.4   Suppose we put a small ice cube in a bucket containing 5 kg of water at 22°C. As the ice melts, it absorbs 35 J of energy from the water. By about how much has the water's entropy decreased?

T8B.5   Suppose the temperature of a 220-g block of aluminum sitting in the sun increases from 18°C to 26°C. By about how much has the water's entropy increased?

T8B.6   Suppose we heat 330 g of water in a pan from 12°C to 92°C. By about how much has its entropy increased?

T8B.7   In a pan on a stove, we convert 125 g of water at 100°C to steam. What is the entropy change of the water that is now steam? (The latent heat of boiling water is 2256 kJ/kg.)

T8B.8   A puddle containing 0.80 kg of water at 0°C freezes on a cold night, becoming ice at 0°C. What was the entropy change of the water that is now ice? (The latent heat of freezing water is 333 J/K.)

Modeling

T8M.1   For a gas with $N = 10^{23}$ molecules, estimate the probability that the expansion into a vacuum shown in figure T8.1 will go in reverse; that is, that all the gas molecules will spontaneously end up in half the container's volume. Express your result in the form "Probability $= 10^{-x}$," where you supply the value of x. (*Hint: Note that $e^y = 10^{y/\ln 10}$.*)

T8M.2   Suppose a flash of laser light adds 28 J of energy to a basin containing 65 kg of water at 18°C. After this energy has distributed itself throughout the water, what is the water's final entropy change? (*Hint: Argue that the water is essentially a reservoir.*)

T8M.3   Suppose we put a 1.0-kg block of aluminum whose initial temperature is 80°C into the ocean at a temperature of 5°C and allow the ocean and block to come into thermal equilibrium.
(a) What was the block's entropy change in this process?
(b) The entropy change of the ocean?

T8M.4   Suppose 22 g of helium gas in a cylinder expands quasistatically while in contact with a reservoir at a temperature of 25°C and does 85 J of work on its surroundings in the process.
(a) What is the gas's entropy change?
(b) What is the reservoir's entropy change?

T8M.5   Suppose we allow 0.40 mol of nitrogen to expand from a volume of 0.005 m³ to a volume of 0.015 m³ while holding the gas at a constant temperature of 304 K. By how much does the entropy of the gas change?

T8M.6   Suppose we put a block of copper with a mass of 320 g and an initial temperature of −35°C into an insulated cup containing 420 g of water at 22°C.
(a) After everything has come to equilibrium, what is the change in the water's and copper's entropy?
(b) Explain why the coating of ice that initially forms around the copper is irrelevant.

T8M.7   Estimate the probability that a 10-g puddle of water will spontaneously freeze on a tabletop at 290 K. Express your result in the form "Probability $= 10^{-x}$," where you supply the value of x. (*Hints: Note that $e^y = 10^{y/\ln 10}$.*)

T8M.8   A plastic bag containing 0.2 kg of water at 20°C is dropped from a height of 0.5 m onto an insulating carpet. Assume that the bag does NOT break. What is the approximate probability that a similar bag of water sitting on a carpet will do the reverse; that is, spontaneously jump 0.5 m in the air? Express your answer in the form "Probability $= 10^{-x}$," where x is a number you will calculate. (*Hint: Note that $e^y = 10^{y/\ln 10}$.*)

T8M.9   Imagine a gas confined in an insulated cylinder by a piston. Suppose that we very suddenly push hard on the piston, compressing the gas to one-half its volume. Then we let the gas slowly expand back to its original volume.
(a) Will the gas have the same temperature as when you started?
(b) Will its entropy be the same as when you started?
(c) Would your answers be different if you had slowly compressed the gas and then slowly expanded it?
Explain all of your responses carefully. (*Hint: When you press the piston in suddenly, do you think that you exert more, less, or the same force on it that you would if you were to compress the gas slowly?*)

T8M.10   A 3-g bullet flying at 420 m/s hits a 2.2-kg aluminum block and embeds itself in the aluminum. After everything has come to equilibrium, by how much has the aluminum block's entropy increased? Be sure to describe the replacement process that you use to actually calculate the entropy change.

Derivations

T8D.1   In this problem, we will see (in outline) how we can calculate the multiplicity of a monatomic ideal gas. This derivation involves concepts presented in chapter T7. Note that the task is to count the number of microstates that are compatible with a given gas macrostate, which we describe by specifying the gas's total energy U (within a tiny range of width dU), the gas's volume V and the number of molecules N in the gas. We will assume that the gas is confined in a cubical container whose sides have length L. We will also assume that the gas is pure (only one type of molecule). Also note that if it is monatomic, its molecules have no accessible internal energy storage modes.

(a) As we saw in chapter T7, the energy of a single molecule in our cubical "box" is

$$E = \frac{h^2}{8mL^2}(n_x^2 + n_y^2 + n_z^2) \tag{T8.35}$$

where m is the molecule's mass, h is Planck's constant, and n_x, n_y, and n_z are independent positive integers. This means that the gas's total thermal energy must be

$$U = \frac{h^2}{8mL^2}(n_{1x}^2 + n_{1y}^2 + n_{1z}^2 + n_{2x}^2 + n_{2y}^2 + n_{2z}^2 + \cdots$$
$$\cdots + n_{Nx}^2 + n_{Ny}^2 + n_{Nz}^2) \tag{T8.36}$$

where N is the number of molecules. Now, imagine a $3N$-dimensional space whose perpendicular axes point in the $1x$, $1y$, $1z$, $2x$, $2y$, $2z$, ..., Nx, Ny, and Nz directions. Each gas microstate corresponds to a complete set of choices of the integers n_{1x} through n_{Nz}, and so to a different tiny cubelet of volume $1^{3N} = 1$ in such a space. The quantity in parentheses is the distance that a given microstate's cubelet is from the origin in that space. In a typical gas, the numbers n_{1x} through n_{Nz} will be large enough that we can consider these quantities to be continuous variables instead of integers. The total number of microstates consistent with the gas's total energy being U within a tiny range $\pm\frac{1}{2}dU$, will be the same as the number of cubelets within a shell of radius $r = \sqrt{n_{1x}^2 + \cdots + n_{Nz}^2} = \sqrt{8mU}\, L/h$ and thickness $dr = \frac{1}{2}(\sqrt{8mU}/L/hU)dU$ in our space. Look up "area of an n-dimensional sphere" online and use what you find to argue that the number of cubelets in such a shell is

$$\Omega = \frac{V^N(8\pi mU/h^2)^{3N/2}}{\Gamma(3N/2)}\frac{dU}{U} \tag{T8.37}$$

where $\Gamma(x)$ is the so-called "gamma function," which is a generalization of the factorial such that $\Gamma(x) = (x-1)!$ when x is an integer, but is a smooth function of x between these values.

(b) But each value of n_{1x} through n_{Nz} must be positive. Argue therefore that the actual number of cubelets in the part of the shell that counts is

$$\Omega = \frac{V^N(2\pi mU/h^2)^{3N/2}}{\Gamma(3N/2)}\frac{dU}{U} \tag{T8.38}$$

(c) Because $x = \frac{3}{2}N$ is ~10^{22} or larger for most everyday gas samples, the distinction between integer and non-integer values of x is negligible. Stirling's approximation (look it up!) states that $x! \approx \sqrt{2\pi x}\,(x/e)^x$ when $x \gg 1$. Use this approximation to argue that

$$\Omega \approx \left(\frac{4\pi meU}{3Nh^2}\right)^{3N/2}\frac{V^N}{\sqrt{3\pi N}}\frac{dU}{U} \tag{T8.39}$$

where I have ignored the distinction between $x - 1$ and x (for $x \sim 10^{22}$, this is an excellent approximation).

(d) However, the molecules in the gas are completely indistinguishable, so microstates that differ only by the rearrangement of values of n_x, n_y, and n_z between different molecules don't actually count as distinct microstates. Therefore, we have overcounted the actual number of microstates by a factor of $N!$ (the number of different ways that we could rearrange the molecules). Use Stirling's approximation again to argue that the correct multiplicity should be

$$\Omega \approx \left(\frac{4\pi me^{5/3}U}{3Nh^2}\right)^{3N/2}\left(\frac{V}{N}\right)^N\frac{dU}{\sqrt{6\pi NU}} \tag{T8.40}$$

(e) Now, ginormous numbers like Ω will not really be changed if we multiply it by even a large number: $(10^{10^{22}})10^{22} = 10^{(10^{22}+22)} \approx 10^{10^{22}}$. So we can drop the factor of $\sqrt{6\pi}$, the factor of dU/U (which will be larger than about 10^{-6} in realistic situations), and even the factor of N). Argue that if we drop these factors, we get the Sackur-Tetrode equation (equation T8.8).

T8D.2   Use the definition of temperature and equation T8.7 to show that $U = \frac{1}{2}fNk_BT$ for an ideal gas.

T8D.3   Consider the following paradox, known as *Gibb's paradox*. Consider the situation shown in figure T8.2, where we divide a gas into equal halves. As stated in the text, we can remove or replace the partition without changing the state of the gas, meaning that removing the partition does *not* increase the gas's entropy. But it should! When we remove the partition, each molecule has double the amount of volume available to it, so the number of microstates available to its should double, meaning that the multiplicity of the gas as a whole *should* go up by a factor of 2^N. Yet it experimentally does not. This is the paradox.

The key to resolving the paradox is to note that this argument does not take into account the fact that gas molecules are indistinguishable even at the microscopic level. Assume that both halves of the gas have equal numbers of particles before the partition is removed. After we remove the partition, some particles in the right half become free to move to the left half and vice versa. Explain why this does *not* actually increase the number of *distinguishable* microstates, even though each molecule's freedom increases.

T8D.4   We saw in figure T8.2 that if we remove a partition separating two identical volumes of a given kind of gas, the entropy does *not* increase. But suppose that the partition separates N molecules of helium gas from N molecules of argon gas, each having the same initial volume V.
(a) Argue on simple physical grounds that the system's entropy *must* increase. (*Hint:* Does the system return to its initial state if we reinsert the partition?)
(b) Use the Sackur-Tetrode equation to calculate the entropy increase. Note that both before and after we remove the partition, the system's total entropy is $S = S_{\text{Hw}} + S_{\text{Ar}}$.

T8D.5   How large is a typical gas's entropy?
(a) Use the Sackur-Tetrode formula to calculate the entropy of 0.1 mole of helium gas at room temperature and atmospheric pressure. (*Hint:* One mole of helium has a mass of 4 g.)
(b) Write the gas's multiplicity in the form $\Omega = 10^{10^x}$ where x is the nearest integer. (*Hint:* Note that $e^y = 10^{y/\ln 10}$.)
(c) Is the multiplicity ginormous or merely large?

T8D.6   The Sackur-Tetrode equation cannot be *quite* right, because at low enough temperatures, it would actually become negative! This is because its derivation *assumes* (see problem T8D.1) that the gas has a high enough energy that the energy levels basically seem continuous and that the gas's multiplicity is ginormous. Suppose that we start with helium gas at a fixed density of 0.165 kg/m³ (about its density at normal atmospheric pressure and temperature) and lower the temperature until the entropy goes negative (indicating that the equation is breaking down).
(a) At what temperature would this happen, assuming that $U = \frac{3}{2}Nk_B T$ (which is also not true at extremely low temperatures), and that helium remains a gas?

(b) The derivation of the Sackur-Tetrode equation also assumes that the gas molecules don't interact. But helium becomes liquid at about 4 K, indicating that the molecules interact enough to make the equation inaccurate at roughly that temperature. Based on your calculation in part (a), is the assumption that molecules don't interact better or worse than other assumptions behind the Sackur-Tetrode equation?

Rich-Context

T8R.1   A high diver dives from the top of a 35-m tower into a pool of water. After the splashing settles down (but before any heat has a chance to be transferred from the water to the diver or vice versa), what is the entropy change of the water in the pool? Explain why one cannot calculate the entropy change for this process directly, and describe the replacement process that you used to calculate the entropy change. Also describe any approximations or estimations that you have to make.

T8R.2   Estimate the probability that a 1-g paper clip sitting on a table at room temperature will spontaneously convert enough of its own or the table's thermal energy to leap 5 cm in the air. Express your result in the form "probability = 10^{-x}," where x is a number that you calculate.

T8R.3   Estimate the entropy of this book, within an order of magnitude or so. (*Hint:* The entropy of nearly any substance at reasonable temperatures is something like Nk_B times the logarithm of something that compares the energy per molecule to some reference energy: see equation T8.8 and equation T3.20. You might calculate this logarithm for a typical monatomic solid or a typical gas. How big or small is the logarithm likely to be? Is it likely to be larger than 100? Smaller than 1/100? Justify your response.)

ANSWERS TO EXERCISES

T8X.1   Consistent: (b), (d), (e); inconsistent: (a), (c).

T8X.2   Since equation T8.28 is based on a rearrangement of equation T8.26, where only temperature *differences* appear, it does not matter whether we use temperatures in degrees Celsius or kelvins (1 K = 1°C of temperature difference). In equation T8.29, on the other hand, we must use absolute temperatures. (I generally try to use absolute temperatures exclusively to reduce my chances of error.)

T8X.3   The block's change in entropy is

$$\Delta S_b = mc \ln\left(\frac{T_f}{T_b}\right)$$

$$= (0.26 \text{ kg})\left(\frac{900 \text{ J}}{\text{kg} \cdot \text{K}}\right) \ln\left(\frac{305 \text{ K}}{362 \text{ K}}\right) = -40 \frac{\text{J}}{\text{K}} \qquad \text{(T8.41)}$$

Heat Engines

Chapter Overview

Introduction

Heat engines play a number of essential roles in our technological society. In this chapter, we explore the limitations that the second law of thermodynamics places on the ability of a heat engine to convert heat energy to useful work.

Section T9.1: Perfect Engines Are Impossible

Designs for a perpetual motion machine that would run endlessly without using fuel fall into two categories: (1) *perpetual motion machines of the first kind*, which violate conservation of energy (the first law of thermodynamics), and (2) *perpetual motion machines of the second kind*, which conserve energy but violate the second law of thermodynamics.

An example of the second category would be a machine that takes heat energy from some thermal reservoir and converts it entirely to useful mechanical energy (work) without doing anything else. During an operating cycle, such an engine would gain entropy along with the heat from the hot reservoir, but since the work it puts out cannot carry away that entropy, the engine's entropy must disappear (in violation of the second law) so that the engine can return to its initial state to begin the next cycle.

Section T9.2: Real Heat Engines

A real heat engine satisfies the second law by exhausting some of the heat it gets from the hot reservoir to a cold reservoir, because this provides a way to carry away the engine's entropy. One can do this and have energy left over to convert to work if there is a temperature difference between the hot and cold reservoirs.

Section T9.3: The Efficiency of a Heat Engine

Since some heat must flow to a cold reservoir to carry away an engine's entropy, no engine can convert all the energy it gets from the heat source to useful work. We define an engine's **efficiency** e to be the fraction of the heat from the hot reservoir that the engine converts to useful work (this is the benefit-to-cost ratio). The second law requires that the net entropy change of the complete system (the hot and cold reservoirs and the engine) not be negative for a cycle of the engine. With a little algebra, one can show that this limits a heat engine's efficiency as follows:

$$e \equiv \frac{|W|}{|Q_H|} \leq \frac{T_H - T_C}{T_H} \qquad \text{(T9.7)}$$

- **Purpose:** This equation describes the limit that the second law of thermodynamics places on the efficiency of a heat engine that taps the heat flow between a hot reservoir at temperature T_H and a cold reservoir at temperature T_C to produce work $|W|$, where $|Q_H|$ is the heat the engine extracts from the hot reservoir.

- **Limitations:** Both T_H and T_C must be absolute temperatures. Some heat engines do not have well-defined hot and cold reservoirs, but if we take T_H to be the highest temperature of the engine's working substance during a cycle, the engine's efficiency will not even approach this limit.

Section T9.4: Consequences

This equation has three important consequences: (1) Any temperature difference can be exploited (in principle) to produce work. (2) The maximum efficiency increases as the temperature difference increases. (3) Waste heat is inevitable. These consequences have a variety of important economic and environmental implications, which are discussed in the section.

Section T9.5: Refrigerators

A refrigerator is essentially a reversed heat engine: it *uses* work to pump heat from a cold reservoir at temperature T_C to a hot reservoir at temperature T_H. We define a refrigerator's **coefficient of performance** (COP) to be the ratio of the heat it removes from the cold reservoir to the work required (again, a benefit-to-cost ratio). A calculation similar to that for a heat engine yields the following limit on a refrigerator's performance:

$$\text{COP} \equiv \frac{|Q_C|}{|W|} \leq \frac{T_C}{T_H - T_C} \qquad (\text{T9.12})$$

- **Purpose:** This equation describes the limit that the second law of thermodynamics imposes on the coefficient of performance of a refrigerator that uses work $|W|$ to extract heat $|Q_C|$ from a cold reservoir at temperature T_C and exhausts energy $|Q_H|$ to a hot reservoir at temperature T_H.
- **Limitations:** Both T_H and T_C must be absolute temperatures. The equation assumes that the refrigerator operates between well-defined reservoirs.

A **heat pump** is a refrigerator that uses the outdoors as a cold reservoir and exhausts its heat into a house. For typical temperature differences, the latter is much larger than the work used, so this is an energy-efficient way to heat a house.

Section T9.6: The Carnot Cycle

The **Carnot cycle** is a sequence of processes using an ideal gas as a working substance that one can use (in principle) to create a maximally efficient heat engine. The sequence is (1) an isothermal expansion with the gas at the same temperature T_H as the hot reservoir, (2) an adiabatic expansion that takes the gas to the temperature T_C of the cold reservoir, (3) an isothermal compression at temperature T_C, and (4) an adiabatic compression that takes the gas temperature back to T_H. The net entropy change of the reservoirs and the gas during this cycle is zero, so an ideal engine based on this cycle would be maximally efficient. One can also create a refrigerator whose COP is maximal by running this cycle in reverse.

However, even a very large Carnot engine operating at anything close to maximal efficiency would generate energy at an impractically small rate. The homework problems discuss engine cycles that are less efficient but more practical.

T9.1   Perfect Engines Are Impossible

Since the dawn of the Industrial Revolution, inventors have been trying to build a **perpetual motion machine** that would run endlessly without using any fuel. (Obviously, such a machine could make its inventor very rich.) Many people have attempted this goal, some quite ingeniously. But all such attempts have failed, and indeed *must* fail. Why?

All designs for perpetual motion machines fall into two broad categories:

1. *Perpetual motion machines of the first kind.* These designs seek to create the energy required for their operation out of nothing, thus violating the law of conservation of energy (the first law of thermodynamics).

2. *Perpetual motion machines of the second kind.* These designs extract the energy required for their operation from sources in a manner that requires the entropy of an isolated system to decrease (thus violating the second law of thermodynamics).

An example of the first kind is a self-powered electric car that uses its motion through the air to turn a windmill that recharges the car's batteries, thus allowing it (in principle) to run forever. But since a moving car always dissipates energy to its surroundings (due to various forms of friction) and since no source for that energy is apparent in this design, it must fail.

Perpetual motion machines of the second kind are usually less obviously flawed. For example, the earth's oceans contain an enormous amount of thermal energy. Why not construct a ship engine that soaks up some of this thermal energy (making the ocean a bit cooler) and converts it to mechanical energy to drive the ship forward? Conservation of energy is not violated here, so it seems more plausible that such an engine might work. Yet such an engine violates the *second* law of thermodynamics. Why?

The key to answering this question lies in understanding that *flowing heat carries entropy from one object to another, but flowing work does not.* Consider first heat energy flowing from a hot object to a cold object. The hot object's entropy decreases in this process, but this is fine because, as we have seen, the cold object's entropy increases at least as much in the process. We can therefore think of this process as carrying the hot object's entropy to the cold object and adding a bit more that was created in the process. But if an object does *work* on another object, this does *not* similarly transfer entropy from the first to the second. For example, if the work energy flowing from the first object raises a weight or adiabatically compresses a gas, it does not change the second object's entropy at all.

Now, figure T9.1a shows an abstract diagram of a "perfect" **heat engine** (such as our hypothetical ocean-powered engine). In a given interval of time (say, one cycle of the engine), such an engine would extract heat energy $|Q_H|$ from a thermal reservoir at temperature T_H (the ocean in our case) and turn that energy entirely into mechanical energy, as shown. The schematic diagram illustrates the flow of energy almost as if it were a river of water. This is a useful mental image: because energy is conserved, it does behave much as a flowing indestructible substance.

Figure T9.1a helps us to understand why such an engine violates the second law of thermodynamics. Imagine that during part of its cycle, the engine absorbs some heat from the ocean. The ocean's entropy decreases during this process, so the engine's entropy must increase at least as much. If, in the next part of the cycle, the engine converts this heat entirely to work, then the energy we got from the ocean flows away, but the entropy remains in the engine. But to repeat the cycle (or maintain its state if it operates in a

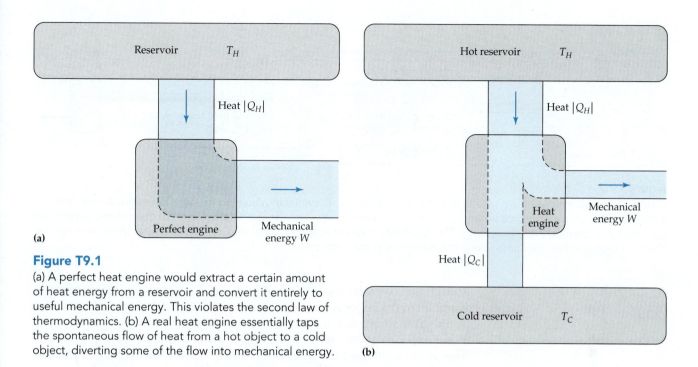

Figure T9.1

(a) A perfect heat engine would extract a certain amount of heat energy from a reservoir and convert it entirely to useful mechanical energy. This violates the second law of thermodynamics. (b) A real heat engine essentially taps the spontaneous flow of heat from a hot object to a cold object, diverting some of the flow into mechanical energy.

steady-state mode), the engine must return to its original macrostate and thus to its original entropy. So the engine must get rid of the entropy it got from the ocean somehow. Since we have no energy left to create a heat flow to carry the entropy away, the entropy must spontaneously disappear, in violation of the second law.

T9.2 Real Heat Engines

So how can we construct a heat engine that satisfies the second law? The trick is as follows. Instead of converting *all* the energy from the heat source to work, the engine needs to save some energy to flow as heat from the engine to something colder, and so carry away the entropy the engine got from the heat source. As we will see, as long as the cold reservoir's temperature T_C (where we dump this waste heat) is lower than the heat source's temperature T_H, we will have a bit of energy left over that we can send out of the engine as work. This process is schematically illustrated in figure T9.1b.

The presence of a *temperature difference* is crucial here. In the language of the financial analogy of chapter T3, the cold reservoir wants the hot reservoir's money (heat), because its happiness (entropy) will increase more in such a transfer than the hot reservoir's happiness decreases. The cold reservoir wants this transfer so much, in fact, that it is willing to pay a broker (the engine) a small profit to mediate the transfer. The broker's happiness is not changed by this transaction (it is just a job), so if the broker charges too much, the cold reservoir does not get enough money so that its happiness increases more than the hot reservoir's decreases, and the deal is off. The maximum amount that the broker can charge is the amount that exactly balances the happiness increase of the cold reservoir with the happiness decrease of the hot reservoir. If there is no generosity difference (temperature difference) between the reservoirs, the broker cannot charge anything.

A good example of a heat engine is a steam turbine in an electric power plant (see figure T9.2). Heat from some source (nuclear power or burning

A visual representation of what a heat engine does

An example of a heat engine: a steam turbine

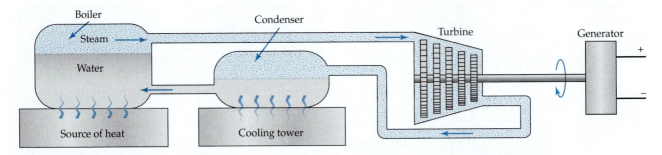

Figure T9.2
A schematic diagram of a steam turbine in an electric power plant. Something (burning fossil fuels, nuclear power) acts as a *hot reservoir*, whose heat boils water, converting it to steam which expands against the blades of the turbine, turning the generator and creating electrical energy. The spent steam is sent to a *condenser* where it gives up heat to the atmosphere (the cold reservoir) and condenses back into water, completing the cycle.

fossil fuels) converts water in a boiler to steam. (Because heat is constantly supplied to the boiler at a constant temperature by whatever the heat source is, the heat source acts as if it were a constant-temperature hot reservoir.) Water expands greatly when it boils, giving the steam a very high pressure. This steam is allowed to expand by blowing against the blades of a turbine, which converts some of the steam's internal energy to mechanical energy (which is ultimately turned into electrical energy by a generator). The spent steam is sent to a condenser, which cools the steam and condenses it back into liquid water by putting it in contact with a cold reservoir. In the bad old days, a nearby river was used as the cold reservoir, but this often increased the temperature of the water enough to be harmful to life downstream. Most modern power plants employ cooling towers, which essentially use the atmosphere as the cold reservoir. In either case, the recondensed water is then returned to the boiler in essentially its original state, completing the cycle.

Any temperature difference can (in principle) be used to drive a heat engine. During the energy crisis of the 1970s, some people looked into the possibility of generating electrical energy from the temperature difference between the ocean surface and ocean floor (which can be as large as 30 K). The plan for such a power plant would be much like that shown in figure T9.2, except that we would have to use a working substance that boils at a lower temperature than water (for example, ammonia). Such a plant would generate electrical power without using any fuel (ultimately, the temperature difference in the ocean is maintained by energy from the sun)! However, some thorny practical and environmental issues make doing this on a large scale impractical. (You can find a good discussion in Penney and Bharathan, "Power from the Sea," *Scientific American*, January 1987.)

T9.3   The Efficiency of a Heat Engine

The purpose of any heat engine is to convert to mechanical energy as much of the heat $|Q_H|$ extracted from the hot reservoir (which is typically energy provided by burning precious fuel) as possible. We can quantify how well an engine does this in terms of its **efficiency** e, which is defined to be the ratio of the mechanical energy $|W|$ produced (the benefit) to the heat energy $|Q_H|$ extracted (the cost):

Definition of efficiency

$$e \equiv \frac{\text{benefit}}{\text{cost}} = \frac{|W|}{|Q_H|} \qquad \text{(T9.1)}$$

Conservation of energy implies that the work $|W|$ that the engine produces is the difference between the energy $|Q_H|$ it gets from the hot reservoir and the energy $|Q_C|$ it sends to the cold reservoir:

$$|W| = |Q_H| - |Q_C| \tag{T9.2}$$

This means that

$$e = \frac{|Q_H| - |Q_C|}{|Q_H|} = 1 - \frac{|Q_C|}{|Q_H|} \tag{T9.3}$$

A perfect engine of the type described in section T9.1 would have $e = 1$: such an engine converts all $|Q_H|$ to W and gives up no heat $|Q_C|$ to a cold reservoir. As we've seen, a perfect engine violates the second law of thermodynamics, so the efficiency of any real engine will therefore be less than 1.

The limit on efficiency imposed by the second law

But how much less? We can calculate this fairly easily. Suppose that one cycle of our heat engine absorbs heat $|Q_H|$ from a hot reservoir at a fixed temperature T_H, produces a certain amount of mechanical energy W, and then gives up heat $|Q_C|$ to a cold reservoir at a fixed temperature T_C. Since both the hot and cold reservoirs are considered to have fixed temperatures, we can calculate the entropy changes of these reservoirs using equation T8.14:

$$\Delta S_H = -\frac{|Q_H|}{T_H} \quad \text{and} \quad \Delta S_C = +\frac{|Q_C|}{T_C} \tag{T9.4}$$

Note that ΔS_H is *negative* because the hot reservoir loses entropy as it gives up heat. The cold reservoir, on the other hand, *gains* entropy as it absorbs heat. In the formula above, I have used the absolute values of the heat lost or gained to make the signs of ΔS_H and ΔS_C explicit.

Exercise T9X.1

The entropy change of the engine itself during a cycle is zero, and the work it creates does not carry away any entropy. Argue that requiring that the entropy of the interacting system consisting of the engine and the reservoirs must increase (or at least remain the same) implies that

$$\frac{|Q_C|}{T_C} - \frac{|Q_H|}{T_H} \geq 0 \tag{T9.5}$$

Exercise T9X.2

Rearrange equation T9.5 to isolate the ratio $|Q_C|/|Q_H|$ and substitute the result into equation T9.3 to show that

$$e \leq \frac{T_H - T_C}{T_H} \tag{T9.6}$$

This inequality applies no matter how cleverly the engine is designed. Any engine that seeks to do better than this will give up too little heat to the cold reservoir to make that reservoir's entropy gain exceed the hot reservoir's entropy loss, and thus it will violate the second law.

Maximum theoretical efficiency of a heat engine

To summarize, we have

$$e \equiv \frac{|W|}{|Q_H|} \leq \frac{T_H - T_C}{T_H} \qquad \text{(T9.7)}$$

- **Purpose:** This equation describes the limit that the second law of thermodynamics places on the efficiency of a heat engine that taps the heat flow between a hot reservoir at temperature T_H and a cold reservoir at temperature T_C to produce work $|W|$, where $|Q_H|$ is the heat the engine extracts from the hot reservoir.
- **Limitations:** Both T_H and T_C must be absolute temperatures. Some heat engines do not have well-defined hot and cold reservoirs, but if we take T_H to be the highest temperature of the engine's working substance during a cycle, the engine's efficiency will not even approach this limit.

Note that this maximum efficiency is the ratio of the temperature *difference* between the reservoirs to the hot reservoir's temperature. As this ratio increases, the maximum possible efficiency goes up. Note also that as the cold reservoir's temperature approaches absolute zero, the efficiency can approach the perfect value of 1.

The derivation of equation T9.7 described here is one of those derivations that every educated person should know and understand (partly because of its economic and environmental consequences, which we will explore shortly). Make *sure* that you understand each step of the derivation and that (in the long run at least) you can reproduce the argument if asked.

T9.4   Consequences

Consequences

As I mentioned before, our technological society extensively employs heat engines in a wide variety of applications. Equation T9.7 therefore has several important conceptual, economic, and environmental consequences.

1. *Any temperature difference can be exploited to generate mechanical energy.* Equation T9.7 tells us that if we have two reservoirs at different temperatures $T_H > T_C$, then in principle we can construct a heat engine that exploits that temperature difference to produce work. One of the reasons we use fossil fuels at such a tremendous rate is that one can relatively easily create a substantial temperature difference by burning such fuels. But this fact also offers some hope that we can avoid using fossil fuels we can exploit *any* temperature difference to convert heat to work.

2. *The greater the temperature difference, the more efficient the engine.* Equation T9.7 makes clear the advantage in making $T_H > T_C$ as large as possible. Because the most practical low-temperature reservoir available to most engines is the surrounding environment (at $T_C \approx 0°C$ to $25°C$), this typically means making T_H as large as possible. The trade-off is usually that engine parts that can withstand extremely high temperatures are expensive, and at a certain point, wasting fuel becomes more economical than using very expensive materials to boost the engine's efficiency by a few percent.

3. *Energy waste is inevitable.* No heat engine can convert all the energy of its fuel to useful mechanical energy: *some* of that energy is necessarily discarded to the cold reservoir. No amount of technological ingenuity will enable one to get around the basic limit on efficiency imposed by the second law of thermodynamics (which by equation T9.7 states).

This last issue has environmental consequences. Disposing of the inevitable waste heat created by large heat engines (or many small engines) in a way that does not detrimentally affect the environment can be challenging. Most cities are substantially warmer than the surrounding countryside partly because of waste heat produced within their borders. People have estimated that the waste heat produced by human activities in the Los Angeles basin currently exceeds 7% of the solar energy falling on the basin.

The problem of disposing of waste heat is particularly acute at large electric power plants, where very large heat engines convert heat to electrical power. Directly dumping the kind of waste heat a power plant produces into a passing river (the most economical method) can have severe environmental consequences. Modern cooling towers operate by taking cold river water, spraying it over the pipes containing the working substance to be cooled, and then allowing the river water to evaporate. Since the latent heat of vaporization of water is fairly large, this carries a lot of heat away for the amount of water vaporized. While this method does not increase the river's temperature, it does reduce the amount of water flowing in the river.

For example, in a typical nuclear power plant, the reactor core produces pressurized steam at about 575 K (one can't make the steam much hotter or the reactor fuel might melt). The temperature of recondensed water leaving the cooling tower might be about 315 K (about 42°C). The maximum *possible* efficiency of the turbines is therefore $e = (575 \text{ K} - 315 \text{ K})/(575 \text{ K}) \approx 0.45$; the efficiency of a real reactor is more like 0.34. So if a plant generates 1000 MW of power (10^9 J/s), it must absorb heat at a rate of $(1000 \text{ MW})/0.34 = 2940$ MW, and therefore give up $2940 \text{ MW} - 1000 \text{ MW} = 1940$ MW as waste heat. This means that nearly two thirds of the energy in the expensive fuel must be wasted! Disposing of waste heat at this rate can significantly heat a large river (see problem T9M.8) or evaporate nearly a ton of water a second in a cooling tower (see problem T9M.9).

Since the efficiency of a heat engine improves as the temperature difference involved increases, heat engines that exploit natural temperature differences (for example, engines that extract power from temperature differences in the oceans, geothermal sources, or created using solar energy) are typically less efficient than fossil-fuel powered engines because natural temperature differences are often small. Though the heat energy provided by such natural sources is essentially "free," naturally driven engines often need to be bigger and more expensive to produce the same amount of power, and disposing of the waste heat can be a big problem (see problem T9M.12).

This does not mean that one cannot extract a lot of energy from such sources. A hurricane is essentially a natural heat engine that gets the motive energy for its winds from relatively small temperature differences between the ocean and the atmosphere (figure T9.3). The conversion efficiency may be small, but the work done by hurricane winds can exceed 10^{17} J/day (which is about the energy released by 1000 Hiroshima-sized atom bombs).

The problem of waste heat

Figure T9.3
A powerful natural heat engine churns up the U. S. East Coast. (Credit: NOAA)

T9.5   Refrigerators

A **refrigerator** is a device that would seem at first glance to violate the second law of thermodynamics. We have seen that heat energy naturally and spontaneously flows from hot to cold because the combined entropy of the hot and cold objects *increases* in this process. The whole purpose of a refrigerator, on the other hand, is to make heat energy flow from cold to hot. How can we do this without violating the second law?

What a refrigerator does

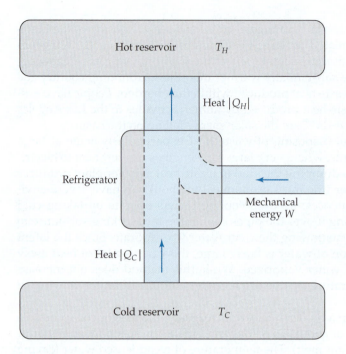

Figure T9.4

A refrigerator is essentially a heat engine operating in reverse: Instead of producing mechanical energy from the natural flow of heat to cold, it uses mechanical energy to drive an unnatural flow of heat from cold to hot.

Obviously, one *can* do this, but one must *supply* work to make this happen. Figure T9.4 makes it clear that a refrigerator is essentially a heat engine operating in reverse (compare with figure T9.1b). Instead of tapping the natural flow of energy from hot to cold to produce a bit of work, a refrigerator uses work to *drive* the unnatural flow of heat flow from cold to hot!

Without this work, a flow of heat energy from cold to hot certainly *would* violate the second law. If energy $|Q_C| = |Q_H|$ were to flow directly from the cold reservoir to the hot reservoir, their entropy changes would be

Second law restrictions on refrigerator performance

$$\Delta S_C = -\frac{|Q_C|}{T_C} \quad \text{and} \quad \Delta S_H = +\frac{|Q_H|}{T_H} \tag{T9.8}$$

respectively. Since $|Q_C| = |Q_H|$ if no work is involved and $T_C < T_H$, this implies that $|\Delta S_C| > |\Delta S_H|$, meaning that the cold reservoir's *loss* of entropy exceeds the hot reservoir's *gain* in entropy, violating the second law. But we can avoid violating the second law by adding enough work so that $|Q_H|$ becomes sufficiently larger than $|Q_C|$ so that

$$\frac{|Q_H|}{T_H} \geq \frac{|Q_C|}{T_C} \quad \text{meaning that} \quad \Delta S_{\text{TOT}} \geq 0 \tag{T9.9}$$

Supplying extra energy to the hot reservoir, therefore, gives it enough extra entropy to make up for the otherwise larger entropy loss of the cold reservoir.

Exercise T9X.3

Multiply both sides of equation T9.9 by T_H and subtract $|Q_C|$ from both sides to show that

$$|Q_H| - |Q_C| \geq \left(\frac{T_H}{T_C} - 1\right)|Q_C| \tag{T9.10}$$

By conservation of energy, $|Q_H| - |Q_C| = |W| =$ the mechanical energy that we must supply, so this equation therefore specifies how large $|W|$ must be.

How can we most usefully quantify a refrigerator's effectiveness? We are most interested in how much heat $|Q_C|$ the refrigerator can extract from the cold reservoir (the benefit) for a given amount of work $|W|$ (the cost). We call their ratios the refrigerator's **coefficient of performance** (COP):

$$\text{COP} \equiv \frac{\text{benefit}}{\text{cost}} = \frac{|Q_C|}{|W|} \qquad \text{(T9.11)}$$

Definition of coefficient of performance (COP)

If we combine equations T9.10 and T9.11, we see that

$$\text{COP} \equiv \frac{|Q_C|}{|W|} \leq \frac{T_C}{T_H - T_C} \qquad \text{(T9.12)}$$

The limit on refrigerator performance

- **Purpose:** This equation describes the limit that the second law of thermodynamics imposes on the coefficient of performance of a refrigerator that uses work $|W|$ to extract heat $|Q_C|$ from a cold reservoir at temperature T_C and exhausts energy $|Q_H|$ to a hot reservoir at temperature T_H.
- **Limitations:** Both T_H and T_C must be absolute temperatures. The equation assumes that the refrigerator operates between well-defined reservoirs.

Exercise T9X.4

Verify that equation T9.12 is correct.

Problem: Estimate the maximum possible coefficient of performance for a standard kitchen refrigerator.

Example T9.1

Solution In the case of an ordinary kitchen refrigerator, the cold reservoir (the inside of the refrigerator) has a temperature of about 40°F $\approx$ 5°C $\approx$ 278 K, while the hot reservoir (the kitchen itself) has a temperature equal to room temperature $\approx$ 295 K, about 17 K warmer. The maximum theoretical COP for such a refrigerator would be about (278 K)/(17 K) = 16.4.

Of course, this calculation ignores the freezer, which has a temperature of about 0°F $\approx$ −18°C $\approx$ 255 K. Many refrigerators actually cool the freezer part directly, and then allow enough heat to flow into the freezer from the rest of the refrigerator to keep the latter cool as well. In such a case, the maximum theoretical COP for such a refrigerator would be (255 K)/(40 K) = 6.4, which is significantly worse.

Note that the COP comes out unitless, which is appropriate for a ratio of energies. Moreover, even though the latter result is not so good, even then such a refrigerator can (in principle) remove an amount of heat from its interior that is more than 6 times larger than the mechanical energy we put in!

Note that COP values will be typically greater than 1! We can take advantage of this fact to construct an effective home heating device.

Example T9.2

Problem: A **heat pump** is essentially a refrigerator that moves heat from outside the house (the cold reservoir) to inside the house (the hot reservoir). Suppose that the outside temperature is 0°C and the inside is 22°C, and that the COP of the heat pump is 8.0. **(a)** Show that this COP is possible. **(b)** If the house requires 36 kW of heat energy to keep its temperature at 22°C, how much energy do we have to supply to the heat pump?

Solution **(a)** Since $T_C = 0°C = 273$ K and $T_H = 22°C = 295$ K, the maximum possible COP for the heat pump is $T_C/(T_H - T_C) = (273\text{ K})/(22\text{ K}) = 12.4$. Therefore, a COP of 8.0 is certainly possible.

(b) The problem states that we must supply an energy $|Q_H| = 36{,}000$ J to the house every second. Since $|Q_H| - |Q_C|$ is equal to the mechanical energy W that we have to supply, and since equation T9.12 implies that $|Q_C| = (COP)W$, we have

$$W = |Q_H| - |Q_C| = |Q_H| - (COP)W \quad \Rightarrow \quad W(1 + COP) = |Q_H| \qquad \text{(T9.13)}$$

Solving for W and substituting in the numbers yields

$$W = \frac{|Q_H|}{1 + COP} = \frac{36\text{ kJ}}{9} = 4\text{ kJ} \qquad \text{(T9.14)}$$

This is the mechanical energy that we must supply each second. Therefore, if we use a heat pump, a mere 4 kJ of mechanical energy will bring 36 kJ of heat energy into the house. This is a real bargain!

If everyone would use heat pumps to heat their houses in the winter, a very large amount of energy could be saved. Unfortunately, heat pumps are quite a bit more expensive (and somewhat less reliable) than standard furnaces (and one can relatively inexpensively heat a home with natural gas, even if supplying 36 kJ of heat requires 36 kJ-worth of natural gas). This makes heat pumps less economically feasible than they should be, considering the value of conserving energy and producing less greenhouse gases.

T9.6   The Carnot Cycle

The **Carnot cycle,** which was first described by the French physicist Sadi Carnot (pronounced *car-NOH*) in 1824, is a hypothetical cyclic process that can be used to convert heat to mechanical energy by using an ideal gas as the working substance. While this particular cycle is not used in realistic engines (for a variety of reasons), it does describe an idealized sequence of gas processes that could be used in principle to produce mechanical energy at the maximum theoretical efficiency allowed by the second law of thermodynamics.

Steps in the Carnot cycle

The Carnot cycle uses an ideal gas confined in a cylinder by a piston, as shown in figure T9.5. Imagine that a flywheel keeps the piston moving in and out of the cylinder at a regular pace and stores the mechanical energy produced by the engine. The Carnot cycle consists of the following four steps:

1. Just as the flywheel begins to pull the piston out, we put the cylinder in thermal contact with a reservoir at temperature T_H, which keeps the gas temperature fixed at T_H. Since the thermal energy U of an ideal gas depends on N and T but not on V, this means that U also is fixed.

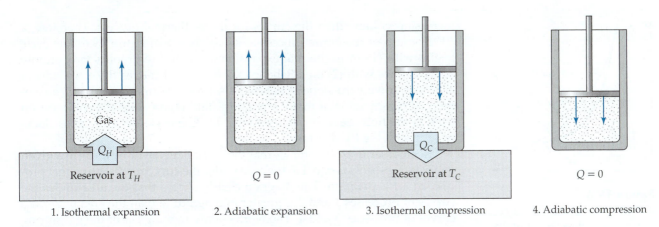

Figure T9.5
The Carnot cycle. You should imagine that the piston is connected to a flywheel that moves the piston in and out and stores the mechanical energy produced by the cycle.

However, the expanding gas does work on the piston, so to keep U fixed, heat must flow into the gas from the reservoir. Let this amount of heat be $|Q_H|$.

2. Now we remove the reservoir. The gas expands adiabatically as the piston continues to move out. Work energy continues to flow out of the gas as it expands, but no heat is coming in to replace it, so the gas's thermal energy, and thus its temperature, decreases. The expansion continues until the gas temperature falls to T_C and the flywheel has pulled the piston to its maximum outward position.

3. The flywheel now begins to push the piston *into* the cylinder while the gas is in contact with the cold reservoir at temperature T_C. Because work energy flows *into* the gas as it is compressed, but the gas's temperature (and thus its thermal energy) does *not* increase, an amount of heat energy equal to the work energy flowing in must flow out of the gas into the cold reservoir. Let us call this amount of heat $|Q_C|$.

4. Finally, we remove the reservoir. The piston continues to compress the gas (now adiabatically). Now work energy flows into the gas, but no heat flows out, so the thermal energy and temperature of the gas increase. The compression continues until the gas temperature reaches T_H, and the piston has moved to its innermost position. The gas is now in the same state as when the cycle started.

In a certain sense, we are using the gas in this process to carry heat from the hot reservoir to the cold reservoir. However, there is a net flow of mechanical energy *out* of the gas: the gas converts *some* of the heat it carries to the cold reservoir to mechanical energy. How do we know? Consider plotting the process on a PV diagram. During the isothermal volume changes (steps 1 and 3), the ideal gas law implies that since $PV = Nk_BT$ and N and T are constant, we have $P \propto 1/V$. On the other hand, during the adiabatic processes (steps 2 and 4), we have

$$PV^\gamma = \text{constant} \quad \text{where} \quad \gamma = 1 + \frac{2}{f} \qquad \text{(T9.15)}$$

where f is the number of degrees of freedom available to each gas molecule (f is 3 for a monatomic gas and 5 for a diatomic gas). This equation implies that $P \propto V^{-\gamma}$, and since $\gamma > 1$, this means that P falls off more sharply with

Why this cycle produces net mechanical energy

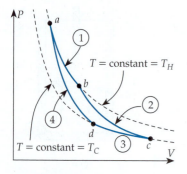

Figure T9.6

A *PV* diagram of the Carnot cycle. The net work flowing out of the gas in the cycle is equal to the area enclosed by the process on the diagram.

The efficiency of a Carnot engine

increasing volume than the curves for the isothermal process. Therefore, a *PV* diagram of the entire process looks as shown in figure T9.6. The cycle begins with (1) an isothermal expansion at the hot reservoir's temperature T_H, continues with (2) an adiabatic expansion that moves the gas's state at a steeper downward slope to the cold reservoir's temperature T_C, (3) an isothermal compression at that temperature, and (4) an adiabatic compression back to the hot reservoir's temperature T_H. The cycle therefore goes clockwise around the *PV* diagram.

Remember from chapter T7 that the work done on or by a gas in a given quasistatic process is equal to the area under the curve representing that process on a *PV* diagram. The diagram clearly shows that the area under the curves of processes 1 and 2 is greater in magnitude than the area under the curves of processes 3 and 4. The work energy moving *out* of the gas during the expansion processes is thus greater than that moving into the gas during the compression processes. So during the cycle as a whole, work energy leaves the gas (and is stored in the flywheel). This must come at the expense of some part of the heat energy that it absorbed in the first step.

One of the main reasons that we are interested in the Carnot cycle is that such an engine (in the absence of friction or other imperfections) would operate at the maximum efficiency allowed by the second law. One can show directly from formulas we derived in chapter T7 for the work done in isothermal and adiabatic processes that $e = (T_H - T_C)/T_H$ for a Carnot engine (see problem T9D.3). But we can arrive at the same result more quickly by considering the entropies involved. During the adiabatic steps, the reservoirs are disconnected, and we know the gas's entropy does not change during an adiabatic volume change, so no new entropy is created in either of these steps. During either of the isothermal steps, the gas and the reservoir have the same temperature, so a heat flow between them also does not create any new entropy (the one gains as much entropy as the other loses). So the net change in entropy of the system consisting of the reservoirs and the engine is *zero* for a complete cycle, implying that the equalities in equations T9.5 and T9.6 apply. This in turn implies that the Carnot engine's efficiency is the maximum allowed by the second law.

One can create a Carnot refrigerator simply by running the Carnot cycle in reverse, that is, by compressing the gas during steps 1 and 2 and expanding it in steps 3 and 4 (that is, running the cycle in the *counterclockwise* direction on the diagram). Now work flows *into* the gas during steps 1 and 2 and out in steps 3 and 4, and we can see that more work flows in than goes out, so the complete cycle converts work to heat going into the hot reservoir, but also extracts heat from the cold reservoir. Since (again) none of the steps creates any new entropy, this refrigerator must have the maximum COP allowed by the second law.

While the Carnot cycle is maximally *efficient*, it is not very practical. To keep from generating new entropy in the isothermal steps, the temperature difference between the gas and the reservoirs must be as small as possible. But the smaller the temperature difference, the slower the heat will flow, so it takes a very long time to get much energy out of such an engine. More realistic engines make compromises between efficiency and power (work per time) generated, which is one reason that realistic engines are less than maximally efficient. Some more realistic engines are discussed in the problems.

TWO-MINUTE PROBLEMS

T9T.1 Classify the following hypothetical perpetual motion machines as being perpetual motion machines of the first kind (A) or the second kind (B).
(a) An electric car runs off a battery which drives the front wheels. The car's rear wheels drive a generator which recharges the battery.
(b) An electric car runs off a battery. When the driver wants to slow down the car, instead of applying the brakes he or she throws a switch that connects the wheels to a generator that converts the car's kinetic energy back to energy in the battery.
(c) An engine's tank is filled with water. When the engine operates, it slowly freezes the water in the tank, converting the energy released to mechanical energy.
(d) Compressed air from a tank blows on a windmill. The windmill is connected to a generator that produces electrical energy. Part of that electrical energy is used to compress more air into the tank.
(e) A normal heat engine is used to drive an electric generator. Part of the power from this generator runs a refrigerator that absorbs the waste heat from the engine and pumps it back into the hot reservoir.

T9T.2 Which of the following devices are heat engines? In each case, answer T (True) if it is a heat engine, and F (False) if it is not.
(a) Your body
(b) An automobile engine
(c) A tornado
(d) The sun
(e) A rocket
(f) A photovoltaic solar panel

T9T.3 In a maximally efficient heat engine, the amount of entropy that the hot reservoir loses as heat flows out of it is exactly balanced by the entropy that the cold reservoir gains as the waste heat flows into it. T or F?

T9T.4 A heat engine produces 300 W of mechanical power while discarding 1200 W into the environment (its cold reservoir). What is this engine's efficiency?
A. 0.20
B. 0.25
C. 0.33
D. Other (specify)

T9T.5 Suppose that, in Iceland, scientists discover geothermal vents that produce pressurized steam at 300°C. Engineers construct a heat engine that uses this steam as a hot reservoir and a nearby glacier as a cold reservoir. What is the maximum possible efficiency of this engine?
A. 300%
B. 100%
C. 52%
D. 45%
E. 22%
F. Other

T9T.6 Suppose you are trying to design a personal fan that you wear on your head and operates between your body temperature (37°C) and room temperature (22°C). What is the maximum possible efficiency of this device?
A. 170%
B. 95%
C. 54%
D. 46%
E. 5%
F. Other (specify)

T9T.7 You can keep your kitchen cool for hours by leaving the freezer door open. T or F?

T9T.8 To function, a home air conditioning system must have access to the home's exterior. T or F?

T9T.9 As the temperature difference between its reservoirs increases, the COP of a refrigerator
A. gets larger.
B. gets smaller.
C. does not change.

T9T.10 A refrigerator uses 100 W of electric power and discards 600 W of thermal power into the kitchen. What is its coefficient of performance?
A. 0.17
B. 0.20
C. 5
D. 6
E. Other (specify)
F. Impossible because it violates conservation of energy.

T9T.11 Someone comes to your house selling a device that draws heat from groundwater and supplies that heat to your house. The salesperson claims that the amount of heat energy entering the house will far exceed the electrical energy supplied to the device. What the salesperson claims here is physically impossible. T or F?

T9T.12 One can convert work to heat (the reverse of what a heat engine does) with perfect efficiency. T or F?

T9T.13 The Carnot engine is important because
A. it was historically the first practical heat engine.
B. it shows (in principle) how one can use ideal gas processes to create a perfectly efficient engine.
C. it shows (in principle) how one can use ideal gas processes to create a maximally efficient engine.
D. it is the only engine that one can run in reverse as a refrigerator.
E. some other reason (specify).

T9T.14 A maximally efficient Carnot engine would produce zero power. T or F?

HOMEWORK PROBLEMS

Basic Skills

T9B.1 We have seen that heat energy cannot be entirely converted to mechanical energy (work). Can mechanical energy be entirely converted to heat? If so, give some examples. If not, explain why not.

T9B.2 A heat engine operates between a hot reservoir at 950°C and a cold reservoir at roughly room temperature (22°C). What is its maximum possible efficiency?

T9B.3 An engine uses water at 100°C and at 0°C as hot and cold reservoirs, respectively. What is its maximum possible efficiency?

T9B.4 A certain heat engine extracts heat energy at a rate of 600 W from a hot reservoir, and it discards energy at a rate of 450 W to its cold reservoir. What is its efficiency?

T9B.5 Why is it important that an air conditioner (which is just a kind of refrigerator) be placed in a window or otherwise have access to the hot environment outside? Would not the air conditioner be more effective if it were placed in the center of the room?

T9B.6 Can you cool your kitchen by leaving the refrigerator door open? Can you heat your kitchen by leaving the oven door open? Explain.

T9B.7 A certain air conditioner maintains the inside of a room at 20°C (68°F) when the temperature outside is 37°C (99°F). What is its maximum possible COP?

T9B.8 A refrigerator (with no freezer) with a COP of 8.4 uses 300 W of electric power to extract heat energy from the interior of the refrigerator, which is at 40°F (5°C). At what rate is heat energy removed from its interior?

Modeling

T9M.1 A hydroelectric power plant, which converts the gravitational potential energy of water stored behind a dam to electrical energy, can operate at very nearly 100% efficiency. How can it do this when a nuclear power plant cannot operate at better than about 40% efficiency?

T9M.2 Imagine that your local power company claims that "heating your home with *electricity* is 100% efficient." In what sense is this true? In what sense is it misleading?

T9M.3 The following wry versions of the first and second laws of thermodynamics have circulated in the physics community for a long time:
> The first law: You can't win.
> The second law: You can't even break even.

Explain in your own words what these restated laws mean.

T9M.4 An old friend comes to you, asking you to invest in the production of a great new toy: a ball that bounces higher with each succeeding bounce. The ball is made of an amazing new kind of synthetic rubber that converts some of its thermal energy to kinetic energy. Explain why you should decline this offer.

T9M.5 The refrigerator COP does *not* express the benefit-to-cost ratio for a heat pump used to heat a building.
(a) What ratio does?
(b) Use an argument similar to that leading from equations T9.8 to T9.12 to determine the limit that the second law imposes on this ratio.

T9M.6 The temperatures generated in the cylinder of an automobile engine can be in excess of 1500 K. Estimate the theoretical maximum possible efficiency of such an engine. (*Hint:* What is T_C going to be, roughly?)

T9M.7 Imagine a power plant designed to exploit the temperature difference between the ocean surface ($\approx$30°C) and the ocean floor ($\approx$4°C). Show that a power plant producing 1000 MW of power would have to dump more than 10,000 MW of waste energy into the cold ocean water.

T9M.8 Suppose that the nuclear power plant described in section T9.4 is located on the banks of a major river that is 67 m wide near the plant, an average of 3 m deep, and that flows at a rate of 0.5 m/s. Suppose that we re-route *all* of this water into the plant and distribute the plant's waste heat evenly throughout the water before returning it to the river bed. How much warmer is the river downstream of the plant compared to upstream? (*Hints:* Water's specific heat is 4.2 kJ·kg^{-1}K^{-1} and its density is 1000 kg/m^3.)

T9M.9 Consider the nuclear power plant described in section T9.4. Suppose that we route water from a nearby river into cooling towers. Suppose that the river near the plant is 67 m wide, an average of 3 m deep, and flows at a rate of 0.5 m/s. About what fraction of the river must evaporate in the cooling towers to carry away the plant's waste heat? (*Hints:* Water's latent heat of vaporization is 2257 kJ/kg and its density is 1000 kg/m^3.)

T9M.10 A freezer with a coefficient of performance of 4.9 uses 250 W of power. How long would it take such a freezer to freeze 15 kg of water?

T9M.11 Suppose that people discover a site in Iceland where temperatures a relatively short distance below the surface of the earth are about 600°C (due to the close proximity of molten rock upwelling from deep inside the earth). A geothermal power plant is constructed at this site. The cold reservoir for this plant is a pool of water constantly fed with ice from a glacier. If the plant produces 100 MW of electrical energy, at what rate is the ice melted? (*Hint:* The latent heat of melting ice is 333 kJ/kg.)

T9M.12 Imagine a solar electric power plant that uses a huge mirror to concentrate light on the boiler for a steam engine. Assume that the boiler generates steam at a temperature of 550°C. The only possibility for a cold-temperature reservoir is a nearby creek that has an average width of 4 m and an average depth of 0.7 m, flowing at a rate of 0.7 m/s. What is the maximum electric power that could be produced by the plant if it boils the entire creek dry?

np to cook dinner! You set up a
its theoretical maximum COP,
m the kitchen at 23°C to supply
ling soup.
ergy must the heat pump use?
(if any) than simply using an

flows through the walls of a
portional to the temperature
ssume also that the air con-
ion of its maximum possible
the cost of air conditioning
oughly as the *square* of that
n carefully.

uld move a certain amount
o a hot reservoir without
n obvious violation of the
). Using energy flow dia-
s T9.1 and T9.4, argue that
you could combine it with
perfect refrigerator.

ease a heat engine's effi-
, should we increase the
decrease the cold reser-
amount? Explain.

lculate the efficiency of a
entropy.
sion in step 1 of the Car-
ture of the gas remains
ws out of the gas as it
he heat energy flowing
ion T7.10 for the work
n to show that

(T9.16)

r step 3 we have

(T9.17)

... with signs. Remember that the absolute values imply that $|Q_H|$ and $|Q_C|$ must be positive, and also remember that $\ln(1/x) = -\ln x$.)

(c) Now consider the adiabatic processes. Argue that since $TV^{\gamma-1}$ is constant for an adiabatic process, we have

$$T_H V_b^{\gamma-1} = T_C V_c^{\gamma-1} \quad \text{and} \quad T_H V_a^{\gamma-1} = T_C V_d^{\gamma-1} \tag{T9.18}$$

(d) Show that this implies that

$$\frac{V_c}{V_d} = \frac{V_b}{V_a} \tag{T9.19}$$

(e) Finally, combine equations T9.16, T9.17, and T9.19 to determine the efficiency of the Carnot engine.

T9D.4 The problem with the maximally efficient Carnot engine design discussed in the text is that, during the isothermal steps, the temperature difference between the gas and the reservoir with which it is in contact is zero. But the rate at which heat flows between the reservoir and gas is typically proportional to the temperature difference between the gas and reservoir, so as the temperature difference goes to zero, the rate at which energy enters or leaves the gas must also go to zero! This means that a maximally efficient Carnot engine will produce zero power.

However, we can get *some* power out of a Carnot engine if we are willing to let the gas's temperature during the first isothermal step be $T_{HG} < T_H$ (the hot reservoir's temperature) and the gas's temperature during the second isothermal step 3 be $T_{CG} > T_C$ (the cold reservoir's temperature). Suppose that both steps take the same time Δt and that the heat that flows during each step is

$$|Q_H| = K(T_H - T_{HG})\Delta t \quad \text{and} \quad |Q_C| = K(T_{CG} - T_C)\Delta t \tag{T9.20}$$

respectively, where the constant of proportionality K is the same for both steps.

(a) For the engine to run steadily, the gas must return to its original state at the end of one complete cycle, meaning that its net entropy change after one complete cycle must be zero. From this, prove both equalities below:

$$\frac{T_H}{T_{HG}} + \frac{T_C}{T_{CG}} = 2 \;\Rightarrow\; T_{CG} = \frac{T_C T_{HG}}{2T_{HG} - T_H} \tag{T9.21}$$

(b) Suppose that the adiabatic steps are essentially instantaneous. Argue that the power (energy per time) produced by this engine is $P = \frac{1}{2}K(T_H - T_{HG} - T_{CG} + T_C)$.

(c) Substitute equation T9.21 into this equation to eliminate T_{CG} and then show that the engine produces its maximum *power* when (with T_H and T_C fixed)

$$0 = 4T_{HG}^2 - 4T_H T_{HG} + T_H^2 - T_C T_H \tag{T9.22}$$

(d) Solve the quadratic to show that the engine produces its maximum power when

$$T_{HG} = \tfrac{1}{2}T_H \pm \tfrac{1}{2}\sqrt{T_C T_H} \tag{T9.23}$$

(e) Find the corresponding value of T_{CG}, and explain why we must choose the plus solution in equation T9.23.

(f) Show that the engine's efficiency at maximum power is

$$e = 1 - \sqrt{\frac{T_C}{T_H}} \tag{T9.24}$$

(*Hint:* Eliminate any radicals in the denominator.)

(g) Compare this efficiency to the maximum Carnot efficiency when $T_H = 600$ K and $T_C = 300$ K.

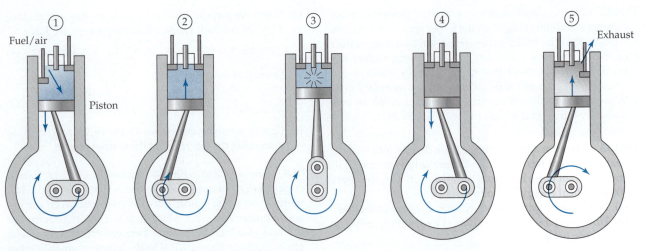

Figure T9.7
The five steps in an automobile engine cycle.

T9D.5 Figure T9.7 illustrates the operation of an idealized automobile engine. We can consider each cycle of the engine to be divided into five steps.

1. The piston moves down the cylinder, drawing in a mixture of gasoline and air through an opened valve at the cylinder's top. We call this step the *intake stroke*.
2. After the valve closes, the piston moves back up the cylinder, compressing the gasoline/air mixture roughly adiabatically. This is the *compression stroke*.
3. When the gas is fully compressed, a spark from the spark plug at the cylinder's top causes the gasoline/air mixture to explode. (In a diesel engine, the increase in temperature caused by the adiabatic compression itself ignites the gas; no spark plug is necessary.)
4. The hot gases produced by the explosion drive the piston down (this is the *power stroke*). The exhaust gases expand adiabatically during this step.
5. Finally, the upward-moving piston pushes the exhaust gases out of the cylinder through an opened valve at its top (this is the *exhaust stroke*). This returns the engine to its original state.

Here is an idealized *PV* diagram for this cycle.

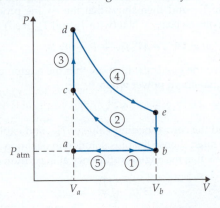

During the intake stroke (step 1), the volume of the cylinder changes from V_a to V_b, drawing in the gasoline/air mixture at atmospheric pressure and ambient temperature.

At point b, the valve closes and the gas is adiabatically compressed to volume V_a again (point c). The gasoline is then ignited, and the temperature of the gas (and thus its pressure) suddenly increases while the piston is almost at rest. The power stroke (step 4) involves an adiabatic expansion of the exhaust gases back to volume V_b. At point e, the exhaust valve is opened, and the pressure in the cylinder suddenly drops to atmospheric pressure, taking the exhaust gas back to point b again. (The explosive decompression during this step is why gasoline engines are noisy and require mufflers.) Finally, the exhaust gas is pushed out of the cylinder at atmospheric pressure during the exhaust stroke (step 5). Note that the number of molecules N in the cylinder changes during steps 1 and 5 but is basically constant during the other steps.

(a) Note that heat energy enters the system during step 3 (from the burning gasoline) and leaves the system in step 5 (carried away by the exhaust). When the cylinder is finally sealed at the end of step 1, the gas at temperature T_e at the end of step 4 has been replaced by an equal amount of gas at temperature T_b, so the net effect of steps 1 and 5 is the same as if we had not let any gas escape but rather cooled it from temperature T_e to temperature T_b. Argue, therefore, that the heat flowing into and out of the engine is

$$|Q_H| = \tfrac{5}{2}Nk_B\,(T_d - T_c) \text{ and } |Q_C| = \tfrac{5}{2}Nk_B\,(T_e - T_b) \quad \text{(T9.25)}$$

(b) Use this to show that the engine's efficiency is

$$e = 1 - \frac{T_e - T_b}{T_d - T_c} \quad \text{(T9.26)}$$

(c) Use the fact that $TV^{\gamma-1}$ during an adiabatic process to show that

$$\frac{T_e - T_b}{T_d - T_c} = \left(\frac{V_a}{V_b}\right)^{2/5} \quad \text{(T9.27)}$$

(d) Find an automobile's efficiency if its compression ratio $V_b/V_a = 8$.

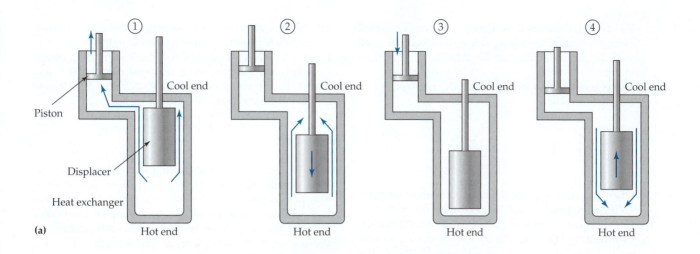

(a)

(b)

Figure T9.8
(a) A schematic diagram of a Stirling engine cycle.
(b) A photograph of an actual toy Stirling engine.
(Credit: © David Haley and Thomas Moore)

Rich-Context

T9R.1   Is it more economical to use a heat pump to heat your home during the winter in California or in Minnesota? Assume that energy costs and house designs are the same in both locations. Also assume that the rate at which a house loses energy is proportional to the temperature difference between the outside and inside of the house.

T9R.2   The *Stirling engine* is a practical heat engine (invented in 1816 by Robert Stirling) widely used in the 19th century before being displaced by the internal combustion engine. It is ingeniously simple, involving few moving parts. Although internal combustion engines generate much more power for a given working volume, Stirling engines are still used in applications where their robust reliability and ability to use any external heat source are valuable.

Figure T9.8a illustrates the Stirling cycle and figure T9.8b shows an actual toy Stirling engine (the fins help keep the cold end of the heat exchanger cool). The piston

and the displacer are both connected to a flywheel in such a way that when the piston is at rest, the displacer is moving, and vice versa. In the first step, air at the hot end of the heat exchanger absorbs heat from the external heat source and expands, pushing around the displacer and against the piston, driving it out. In step 2, the piston comes essentially to rest, and the displacer moves toward the hot end of the exchanger, displacing the air to the cold end of the exchanger. In step 3, the flywheel begins to push the piston in, but now most of the air inside the engine is at the cool end of the heat exchanger, and so it takes less work to compress it than we got out during step 1. Finally, the piston comes to rest again, and the displacer moves toward the cold end, displacing the air back to the hot end, and we are back where we started.

(a)   For the sake of simplicity, assume that steps 2 and 4 occur at exactly constant volumes and that all the air has plenty of time to cool to the temperature T_C of the cool end as the displacer moves during step 2 and to warm to the temperature T_H of the hot end as the displacer moves during step 4. Ignore the small amount of air on the wrong side of the displacer in all steps. Draw a *PV* diagram for this idealized Stirling cycle.

(b) Use equation T7.10 to argue that the net work done during this cycle is

$$|W| = Nk_B (T_H - T_C) \ln\left(\frac{V_{max}}{V_{min}}\right) \tag{T9.28}$$

where V_{max} and V_{min} are the maximum and minimum volumes, respectively, of the gas during the cycle.

(c) Heat flows into the gas in both steps 1 and 4. What is the total heat $|Q_H|$ flowing into the engine during a cycle, expressed in terms of the same quantities? (Remember that air is diatomic.)

(d) Show that the Stirling engine's efficiency e is such that

$$\frac{1}{e} = \frac{1}{e_C} + f\left(\frac{V_{max}}{V_{min}}\right) \tag{T9.29}$$

where $e_C = (T_H - T_C)/T_H$ and $f(V_{max}/V_{min})$ is a function of V_{max}/V_{min} that you should determine.

(e) Argue that the efficiency of even our idealized Stirling engine is less than that of a Carnot engine.

(f) What is the efficiency of our idealized Stirling engine when operating between temperatures of 600 K and 300 K, if its compression ratio $V_{max}/V_{min} \approx 3$? Compare this to the efficiency of a Carnot engine.

Advanced

T9A.1　Imagine that you are trying to construct a practical Carnot engine that will deliver 1 horsepower (hp) $\approx$ 700 W. The parameters are as follows. For safety reasons, your high-temperature reservoir can be no higher than 535°F (280°C). Your engine is cooled by the surrounding air. Because we must allow plenty of time for heat to flow and volume changes need to be reasonably quasistatic, you are limited to 2 cycles per second. The metal used for your cylinder can withstand 25 atm of pressure, and various considerations limit the ratio of your maximum-to-minimum volume to be no more than 10. Find the minimum and maximum volumes for your cylinder. (For comparison, a gasoline engine with a cylinder volume of a few thousand cubic centimeters can produce roughly 100 hp.) (*Hints:* Study problem T9D.3. Note that $|W| = e|Q_H|$ and $|Q_H|$ is given by equation T9.16. This will get you started.)

ANSWERS TO EXERCISES

T9X.1　The second law of thermodynamics requires that the total entropy of the interacting objects here (the two reservoirs and the engine) not decrease:

$$\Delta S_{TOT} = \Delta S_C + \Delta S_{engine} + \Delta S_H \geq 0 \tag{T9.30}$$

As discussed, $\Delta S_{engine} = 0$ (at least in the long run). Substituting the values of ΔS_H and ΔS_C from equation T9.4 yields the desired result.

T9X.2　Multiplying both sides by $T_C|Q_H|$, we get

$$\frac{|Q_C|}{|Q_H|} - \frac{T_C}{T_H} \geq 0 \tag{T9.31}$$

Adding T_C/T_H to both sides gives the desired result.

T9X.3　Equation T9.3 reads

$$e = 1 - \frac{|Q_C|}{|Q_H|} \tag{T9.32}$$

This will be *smaller* than the quantity $1 - T_C/T_H$, because according to equation T9.6, $|Q_C|/|Q_H| \geq T_C/T_H$, so subtracting $|Q_C|/|Q_H|$ from 1 yields a smaller number than subtracting T_C/T_H from 1. Therefore (as stated)

$$e \leq 1 - \frac{T_C}{T_H} \tag{T9.33}$$

T9X.4　Since $W = |Q_H| - |Q_C|$, equation T9.11 implies that

$$COP = \frac{|Q_C|}{W} = \frac{|Q_C|}{|Q_H| - |Q_C|} \tag{T9.34}$$

Plugging equation T9.10 into this (and noting that when we *divide* by a bigger number the result is *smaller*), we get

$$COP \leq \frac{|Q_C|}{(T_H/T_C - 1)|Q_C|} = \left(\frac{T_H}{T_C} - 1\right)^{-1} \tag{T9.35}$$

If we replace 1 by T_C/T_C, add the terms, and invert the ratio, we get equation T9.12.

T10

The Physics of Climate Change

Chapter Overview

Section T10.1: Introduction

In this closing chapter of the unit, we will use the physics that we have learned in this unit to construct a simplistic but conservative model of climate change. This will illustrate not only the process of model-building but also its application to a significant real-world problem.

The fundamental physical facts behind climate change are:

- Energy is conserved.
- An object with a nonzero absolute temperature radiates electromagnetic energy according to the Stefan-Boltzmann law ($P = \epsilon \sigma A T^4$, where P is the power radiates, ϵ is object's emissivity, σ is the Stefan-Boltzmann constant, A is the object's surface area, and T is the object's absolute temperature).
- The peak wavelength of this radiation is given by Wien's law.
- Carbon dioxide (CO_2) is opaque to infrared light.
- CO_2 concentration in the atmosphere is increasing.

In particular, the average atmospheric concentration of CO_2 has been 280 ppm (parts per million) for many millenia before 1900, but was 400 ppm in 2015 and is steadily climbing.

Section T10.2: Radiative Equilibrium

Conservation of energy requires that if the earth's temperature is to remain steady, then the amount of energy that the earth receives from the sun must be exactly balanced by the amount of energy that the earth re-radiates to space. The earth receives energy from the sun in proportion to the cross-sectional area πR^2 that it presents to sunlight (where R is the earth's radius) but radiates from its entire surface area $4\pi R^2$. Also, snow and clouds reflect about 30% of the sun's energy directly back to space. The earth's emissivity is $\epsilon \approx 1$ at the infrared wavelengths that the earth emits. The bottom line is that the earth must thermally radiate an average of about 240 W/m^2 of energy back into space, which requires (by the Stefan-Boltzmann law) that it have an average surface temperature of about $T = 255$ K $= -18°$C, well below the empirical historical global average of 14°C.

Section T10.3: The Atmospheric Blanket

The discrepancy is partly because we have not included the effects of the earth's atmosphere. At the wavelengths that the earth thermally radiates (about 10 μm), the carbon dioxide in the earth's atmosphere is opaque. We can crudely model a given amount of carbon dioxide in the atmosphere as n layers of opaque blankets just above the earth's surface. Each is transparent to sunlight, but each completely absorbs the thermal radiation that the layer below emits and then reradiates half that energy upward and half downward. The energy balance between the earth, sun, the blankets, and space imply that the earth must radiate $n + 1$ times as much energy as it receives directly from the sun, which means (by the Stefan-Boltzmann law) that its temperature must be a factor of $(n + 1)^{1/4}$ higher than 255 K.

Section T10.4: Building a Better Model

To get the pre-1900 observed average global temperature of 287 K = 14°C, we must have $n = 0.605$, meaning that the carbon dioxide in the atmosphere does not comprise a completely opaque layer, but allows a fraction of the earth's thermal radiation to directly escape to space. If we also note that only 27% of the atmosphere's opacity is due to carbon dioxide (the rest is due to water vapor, whose concentration various geophysical cycles keep very constant), then we can predict that at the current concentration of CO_2 in the atmosphere (400 ppm), the earth's equilibrium surface temperature should be about 1.4°C above the baseline 1900 average temperature of 14°C. This is a bit larger than the observed shift of about 0.8°C (as of 2015).

Section T10.5: The Ocean Effect

The remaining discrepancy is primarily due to the fact that the ocean has a such a large heat capacity that it would take about 30 years for it to store enough heat energy for the earth to come into equilibrium at a given CO_2 concentration. The earth's current temperature therefore roughly reflects the temperature that would correspond to the equilibrium temperature when the CO_2 concentration was at the level that it was 30 years ago, or about 345 ppm. If one calculates the temperature for this concentration, one gets a temperature increase of 0.8°C, which is basically spot on.

This means that even if we were to stop producing carbon dioxide in 2016, we are essentially committed to a temperature increase of 1.3°C above the baseline (assuming the 2016 CO_2 concentration of 400 ppm) in 30 years.

Section T10.6: Improving the Model

Just because this extremely crude model agrees with present data does not mean that it will accurately predict the future. Climate scientists are working very hard to make the model more realistic by (1) treating the physics of the atmosphere in a more nuanced way, (2) allowing the temperature to vary (with as fine a resolution as is practical) on different points of the earth's surface and at different heights in the atmosphere, (3) include the effects of secondary effects and feedback loops involving evaporation, cloud cover, snow cover, and so on, and (4) actively modeling dynamic effects of heat transfer and storage in the ocean and atmosphere. For more details, consult the website scienceofdoom.com.

Section T10.7: Consequences

These models can provide useful predictions of some of the consequences of climate change, some of which have already been supported by empirical data. Such consequences include (1) an increase in extreme weather events, (2) sea level rise, (3) more powerful storms, and (4) changes in patterns of rainfall and drought. The section discusses some of the evidence for these consequences and why they make sense.

Section T10.8: Concluding Scientific Postscript

The point is that sound physics models are crucial in exposing this global problem and its consequences. They are also crucial in providing the tools that we need to successfully deal with the problem and its consequences. Indeed science, with its emphasis on imagination, creativity, community, empiricism, and commitment to the truth provides unique and essential tools for solving problems like these. Your knowledge of physics already puts you in the top 1% of people in the world best able to deal with this problem. Be part of the solution!

T10.1   Introduction

You don't have to *believe* in climate change; you can *calculate* it. Probably less than 1% of the population in the United States has the physics background needed to do this calculation, but *you do*. In the next sections, we will construct a simple but very conservative model for predicting the effects of this increase in CO_2 concentration. This provides an excellent closing example not only of the whole process of model-building, but also of why knowing the physics of a situation can be important in the real world.

The basic physical facts behind climate change

The fundamental physics behind climate change is actually pretty simple, and is based on five basic physical facts that are beyond dispute.

- Energy is conserved.
- An object with a nonzero absolute temperature radiates electromagnetic energy according to the Stefan-Boltzmann law.
- The peak wavelength of this radiation is given by Wien's law.
- Carbon dioxide (CO_2) is opaque to infrared light.
- CO_2 concentration in the atmosphere is increasing.

Fleshing out these facts

The Stefan-Boltzmann law (which we have discussed earlier) states that an object at absolute temperature T radiates energy at a rate P such that

$$P = \epsilon \sigma A T^4 \tag{T10.1}$$

where ϵ is the object's emissivity (ranging from nearly 0 for something shiny to 1 for a so-called "perfect blackbody"), A is the area of the emitting object's surface, and σ is the Stefan-Boltzmann constant $= 5.67 \times 10^{-8}$ W·m^{-2}K^{-4}. Wien's law states that though this radiation is spread out over a range of wavelengths, the *peak* wavelength (that is, the wavelength at which the greatest amount of energy is emitted) for any object is given by

$$\lambda = \frac{b}{T} \tag{T10.2}$$

where $b = 2{,}898{,}000$ nm·K. For an object that has roughly the earth's current average temperature of 15°C or 288 K, this wavelength is

$$\lambda = \frac{2{,}898{,}000 \text{ nm·K}}{288 \text{ K}} = 10{,}060 \text{ nm} = 10.06 \text{ μm} \tag{T10.3}$$

which is in the infrared, well below the wavelength range of visible light (700 nm − 400 nm). Experimentally, we find that though CO_2 gas is transparent at visual wavelengths, it's opaque at wavelengths near 10 μm.

Finally, we know that the CO_2 concentration in the atmosphere is increasing. During the whole of human history until recently, this concentration has been very nearly 280 ppm. Figure T10.1 shows the measured concentration of CO_2 during the last few decades. The current concentration (as I write this in 2015) is 400 ppm and climbing (as you can see) at a rate of about 2 ppm per year (and accelerating). This rate of increase roughly corresponds with estimates of the CO_2 emitted by human activity, but from the point of view of the physics, it does not matter where the CO_2 is coming from. That it is increasing at rates unprecedented in millennia is an experimental fact.

T10.2   Radiative Equilibrium

Conservation of energy requires that if the earth's temperature is to remain constant, the thermal energy that flows into the earth must be equal to the

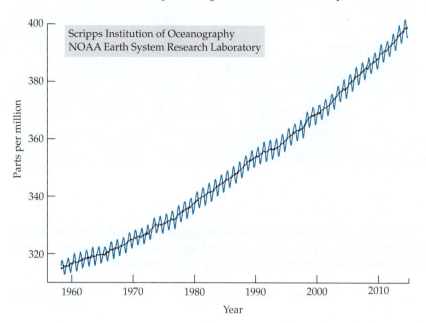

Atmospheric CO₂ at Mauna Loa Observatory

Scripps Institution of Oceanography
NOAA Earth System Research Laboratory

Figure T10.1
A graph showing the carbon dioxide in the atmosphere measured at the Mauna Loa Observatory in Hawaii as of November 2014. The annual variation in this concentration is due to vegetation growth in the northern hemisphere each northern summer (the earth has far less land surface in the southern hemisphere). The black line running through the center is seasonally corrected data. (Credit: NOAA)

thermal energy that flows out. Almost all of the energy that flows into the earth is from the sun (a tiny amount comes from radioactive decay processes in the earth's interior, but this is very small in comparison). Our simple model will assume that *all* of this energy comes from the sun. Above the earth's atmosphere, we measure the intensity of sunlight to be 1367 W/m². Of this, very nearly 30% (on the average) is reflected directly back into space by clouds and snow. Therefore, the area of the earth facing the sun *absorbs* (on average) a power per unit area of 960 W/m² of energy from the sun.

Now the cross-sectional area that the earth presents to the sun is πR^2 (where R is the earth's radius), but once the earth absorbs the energy, it re-radiates it into space from its entire surface area of $4\pi R^2$. Therefore, the earth's average temperature T in equilibrium must be such that the power per unit area it radiates back into space is $\frac{1}{4}(960 \text{ W/m}^2) = 240 \text{ W/m}^2$ for the energy outflow to match the energy inflow (as figure T10.2 illustrates).

How much power the earth must radiate to space to remain in thermal equilibrium

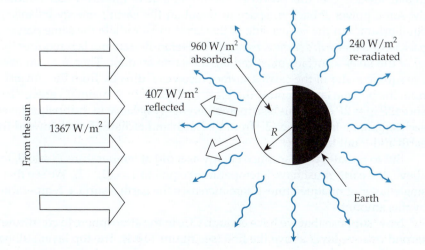

960 W/m²
absorbed

240 W/m²
re-radiated

407 W/m²
reflected

From the sun

1367 W/m²

R

Earth

Figure T10.2
The earth's solar energy balance. The earth's cross-sectional area πR^2 absorbs an average power of 960 W per square meter facing the sun. Since it re-radiates this energy from its entire surface area of $4\pi R^2$, it must radiate $\frac{1}{4}(960 \text{ W})$ or 240 W per square meter to maintain the energy balance.

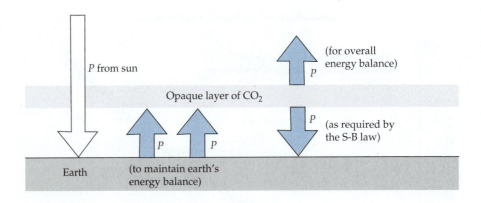

Figure T10.3
A single layer of opaque CO_2 makes the earth warmer, because to keep outgoing energy equal to incoming energy, the earth must radiate *twice* the power it would without an atmosphere, meaning that its temperature must be $2^{1/4}$ times larger than it would be without an atmosphere.

The earth's predicted temperature (with no atmosphere)

The earth's average temperature T in equilibrium will therefore be that required so that the earth's surface radiates an average of 240 W/m². According to the Stefan-Boltzmann law, this temperature must be

$$\frac{P}{A} = \epsilon \sigma T^4 \quad \Rightarrow \quad T = \left(\frac{P}{A}\frac{1}{\epsilon \sigma}\right)^{1/4} \tag{T10.4}$$

Assume that the earth's atmosphere and surface are nearly perfect emitters ($\epsilon \approx 1$) at infrared wavelengths near 10 μm (this is actually an excellent approximation). Then the required average temperature is

$$T_0 = \left(240 \, \frac{W}{m^2} \frac{1}{5.67 \times 10^{-8} \, W \cdot m^{-2} K^{-4}}\right)^{1/4} \approx 255 \text{ K} \tag{T10.5}$$

This is about −18°C. Since the earth's average temperature until quite recently is about +14°C (287 K), we see that our model (so far) is too simplistic.

T10.3   The Atmospheric Blanket

Modeling the atmosphere as one completely opaque layer

This actually *would* be the earth's average temperature if it did not have an atmosphere. But the CO_2 in the atmosphere makes a big difference. To see why, suppose that the atmosphere contained one layer of CO_2 that is entirely opaque to the infrared light that the earth radiates (see figure T10.3). Of course, this layer is transparent to ordinary light, so the sun directly delivers an average power of 240 W to every square meter of the earth's total surface to that surface. Call this power P. The upper side of this layer has to radiate the same power P back to space to maintain the earth's energy balance in equilibrium. But the *bottom* side of this layer must radiate the same power P back *toward the earth's surface*, because the Stefan-Boltzmann law requires that *all* of the layer's surface must radiate the same amount. Therefore, an average square meter of the earth's surface receives P directly from the sun *and* P from the opaque layer. It must therefore re-radiate $2P$ to maintain energy balance ($2P$ in = $2P$ out). This also maintains energy balance in the opaque layer: each square meter receives $2P$ from the earth and radiates P back toward the earth and P out to space.

But to radiate twice as much energy as it did at temperature T_0 calculated above, the earth must have a temperature equal to $T = (2)^{1/4}T_0$. We see that a single layer of opaque atmosphere increases the earth's surface temperature by this amount.

Modeling the atmosphere as n completely opaque layers

Now suppose that we have enough CO_2 in the atmosphere to constitute a second opaque layer above the first (see figure T10.4). The top layer still has

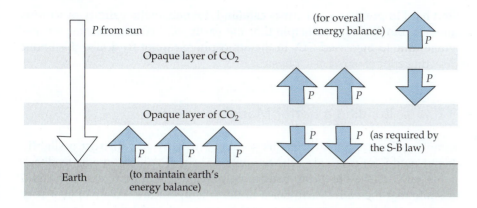

Figure T10.4
When we have two opaque layers of CO_2, energy flows are balanced when the earth emits *three* times the power it would without an atmosphere.

to radiate P to space to balance the net power P that flows into the system from the sun. So it must also radiate P downward. This requires an inflow of $2P$ from the middle layer to keep the top layer balanced. But this middle layer must then also radiate $2P$ toward the earth's surface, which in turn must radiate $3P$ upward to keep the middle layer in balance (and also itself in balance as it is receiving P from the sun and $2P$ from the middle layer). Therefore, with *two* layers of CO_2, the earth's temperature must be a factor of $(3)^{1/4}$ times higher than its temperature T_0 for zero layers.

It is pretty clear that the general formula for n layers is going to be

$$T = (n + 1)^{1/4} T_0 \tag{T10.6}$$

We can see this in action on the planet Venus. At the distance that Venus is from the sun (and noting that its clouds reflect 90% of the sunlight falling on the planet, not 30% as for the earth), its average surface temperature should be 184 K (actually colder than the earth!). But its average surface temperature is a whopping 737 K (see http://nssdc.gsfc.nasa.gov/planetary/factsheet /venusfact.html). This is because Venus's very thick atmosphere, which is almost pure CO_2 and which has a pressure over 90 atmospheres at Venus's surface, amounts to the equivalent of about 256 opaque layers!

Exercise T10X.1

Verify that 256 layers would yield the observed surface temperature for Venus if its average temperature were $T_0 = 184$ K for zero layers.

The point of this section is that if the concentration of CO_2 in the earth's atmosphere goes up (other things remaining the same), the earth *must* get warmer. This is not debatable. It is a direct logical consequence of the five basic principles described in the opening section.

The exact *amount* of the increase, however, is hard to predict. The reason is that the earth is actually a very complicated interacting system, not the uniform ball that I have been assuming. In particular, if the changing temperature leads to changes in the amount of light that the earth reflects, then the actual temperature change will be different than this simple model would predict. For example, scientists now think it probable that arctic ice will disappear in the summer within a decade due to warming temperatures. This would decrease the light reflected into space by the Arctic dramatically, increasing the energy the earth collects from the sun, and thus further increasing the temperature change. Such feedback loops can be very complicated

The point: more CO_2 means a warmer planet

and hard to predict, so scientists can (and do) debate the validity of various models. But the core principle that the earth *will* become warmer (on average) due to increased CO_2 in the atmosphere cannot be doubted without rejecting a lot of very basic physics.

T10.4   Building a Better Model

Granting this complexity, in this section, I will try to construct a slightly better model for predicting the temperature rise. I will assume *no* feedback loops such as the one described in the previous paragraph, and I will still assume that the entire earth has a uniform temperature. This will be a very crude model, but we can use it to make very rough estimates.

How many layers does the pre-industrial concentration of CO_2 (280 ppm) represent?

We saw from equation T10.6 that the earth's average temperature should be $T = (n + 1)^{1/4} T_0$, where n is the number of opaque layers and $T_0 = 255$ K is the earth's temperature with no layers. This formula must be correct for integer values of n, but let's assume that it works for non-integer values as well.

Now, we know that before 1900 the earth's average temperature was a relatively stable 14°C = 287 K ≡ T_b (for "baseline temperature") and that the concentration of CO_2 in the earth's atmosphere was 280 ppm. Let's use this information to calculate the baseline number n_b of opaque layers in the earth's atmosphere. According to equation T10.6, we must have

$$(n_b + 1)^{1/4} = \frac{T_b}{T_0} \quad \Rightarrow \quad n_b = \left(\frac{T_b}{T_0}\right)^4 - 1 = \left(\frac{287 \text{ K}}{255 \text{ K}}\right)^4 - 1 = 0.605 \qquad (T10.7)$$

This tells us that the baseline earth atmosphere did not have even one complete layer, but only a partial layer that allows some of the earth's thermal energy to radiate directly to space.

Predicting the earth's temperature at the current 400 ppm concentration

Given this model, what would we predict the temperature to be at the 2015 CO_2 concentration of 400 ppm? It turns out (for some pretty subtle reasons: see section T10.6) that the effective number of opaque layers is proportional to the *square root* of the concentration. The 2015 value of n should therefore be

$$n = n_b \left(\frac{400 \text{ ppm}}{280 \text{ ppm}}\right)^{1/2} = 0.605 \left(\frac{400}{280}\right)^{1/2} = 0.723 \qquad (T10.8a)$$

and so the 2015 average temperature should be

$$T = (n + 1)^{1/4} T_0 = (1.723)^{1/4} (255 \text{ K}) = 292 \text{ K} \qquad (T10.8b)$$

This is a 5 K = 5°C increase over the 1900 baseline temperature $T_b = 287$ K. This is inconsistent with an actually measured average temperature increase (which is also not really disputed) of about 0.8°C.

Correcting for the fact that CO_2 is only a small part of the blanket

This is because CO_2 actually only contributes about 27% of the atmosphere's opacity (see section T10.6), with water vapor contributing most of the rest. Let's fix this. Since 0.27(0.605) = 0.163, this is the effective thickness of the baseline CO_2 layer and 0.605 − 0.163 = 0.442 is the effective thickness of the rest, which we will assume remains constant. Therefore, the effective opacity of the atmosphere after CO_2 has climbed to 400 ppm is actually

$$n = 0.163 \left(\frac{400}{280}\right)^{1/2} + 0.442 = 0.637$$

$$\Rightarrow T = (n + 1)^{1/4} T_0 = (1.637)^{1/4} (255 \text{ K}) = 288.4 \text{ K} \qquad (T10.9)$$

This is better, but a difference of 1.4°C from the baseline is still a bit high.

T10.5 The Ocean Effect

What could be wrong? This would be the right result if the earth were in equilibrium (that is, if it were radiating as much energy as it receives). Over the long term, the earth's energy flows must balance, but in the short term, the earth can absorb more energy than it radiates.

An analogy that might help explain this is that of a speeding car. In equilibrium, a car's cruising speed is that speed where the energy inflow from the engine exactly balances the energy outflow due to friction (mostly air drag). If you increase the energy flow from the engine, the car will come to a higher cruising speed where the outflow again matches the inflow. The car's speed is therefore analogous to the planet's temperature.

Adding CO_2 to the atmosphere is thus like stomping down on the car's accelerator. But the car's speed will not instantly adjust to the change in energy flow from the engine. Rather, some of the engine's energy must be used to increase the car's kinetic energy, since the car will have a higher kinetic energy at the final new cruising speed. Since the extra energy is flowing from the engine at a fixed rate, it takes some time to ramp up the car's speed.

Similarly, it takes time for the earth's surface to ramp up its internal energy (and thus its temperature) in response to the net increase in energy that is flowing into it. It turns out that because water has a very high specific heat and because currents mix ocean water continually, the ocean can absorb an enormous amount of energy without its surface temperature changing very much. (Dry land can also absorb some energy but responds much more quickly, because rock does not conduct heat very well.)

The ocean therefore slows down the rise in the earth's average temperature increase, causing it to lag about 30 years behind the value that would be predicted for equilibrium. Therefore, the average temperature that we are experiencing now actually reflects the CO_2 concentration in the air *30 years ago*. As we see from figure T10.5, this was about 345 ppm.

The ocean delays equilibrium by about 30 years

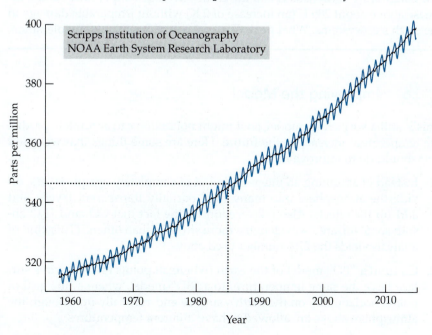

Atmospheric CO₂ at Mauna Loa Observatory

Scripps Institution of Oceanography
NOAA Earth System Research Laboratory

Figure T10.5
This graph shows that the concentration of CO_2 at a date 30 years before 2015 about 345 ppm. (Credit: NOAA)

The predicted temperature
accounting for this delay

Including this delay, the earth's current average temperature should be

$$n = 0.163\left(\frac{345}{280}\right)^{1/2} + 0.442 = 0.623$$

$$\Rightarrow T = (n+1)^{1/4}T_0 = (1.623)^{1/4}(255 \text{ K}) = 287.8 \text{ K} \qquad \text{(T10.10)}$$

which is almost exactly what we observe!

One of the implications of this model is that even if we were to completely stop emitting CO_2 into the atmosphere *today*, we would still be *committed* to the temperature increase required for the earth to come into equilibrium with 400 ppm of CO_2 in the atmosphere, or 1.4°C (as calculated in equation T10.9). And even if we begin actively and artificially *reducing* the CO_2 in the atmosphere, it will take nearly a generation after we begin to see any payoff. The earth's temperature for the coming decade has actually been determined by decisions that people have made *before you were even born*.

Postscript: One might wonder why, since water vapor contributes most of the opacity in the atmosphere, we aren't worried about human emission of steam. The answer is that the concentration of water vapor in the atmosphere is maintained by active geologic processes that make sure that any excess water injected into the atmosphere condenses out and the equilibrium concentration is restored very rapidly, on the time scale of days. Since CO_2 is much more inert, geologic processes that remove excess CO_2 from the atmosphere have much longer time scales, more like 1000 years. Therefore, any *water* we add to the atmosphere is quickly removed, but any CO_2 we add to the atmosphere remains there for a very long time on the human scale.

Exercise T10X.2

Assuming that we continue to add CO_2 to the atmosphere at a rate of 2 ppm/y for the near future (corresponding to *zero growth* in fossil fuel use), what does our model predict the earth's average temperature will be in 2050?

Exercise T10X.3

The community consensus is that the earth's average temperature cannot increase above about 289 K (an increase of 2 K) without irreparable damage to the earth's ecosystems. What peak CO_2 concentration is therefore tolerable?

T10.6   Improving the Model

How to make the model
more accurate

This is still a very crude model, so it might not *continue* to accurately predict the temperature increase in the future. Here are some things that we could and should do to improve it:

1. Instead of assuming an integer number of completely opaque layers, divide the atmosphere into many thin, partially transparent layers, and add up the effects. Also take account of the fact that CO_2 and H_2O absorb some infrared wavelengths more strongly than others. (This kind of analysis yields the 27% number used above.)

2. Go from a "0D" model of the earth (where all points on the earth's surface have the same temperature) to a "3D" model, where a fine grid of points horizontally on the earth's surface and vertically up through the atmosphere above are allowed to have different temperatures.

3. Factor in effects of various feedback loops between increased temperature and water vaporization, cloud formation, ice depletion, and so on.

4. Factor in the dynamics of heat being stored and released by the ocean, land, and atmosphere, and similar dynamic effects.

None of these details affect the basic physics outlined earlier, only the prediction of exactly when, where, and how much warming. But scientists are working very hard on models incorporating the improvements listed and more. The steadily increasing power of computers and the large number of independent models that people have constructed help make the community's predictions about climate more precise and objective every year.

In spite of the silly name, "scienceofdoom.com" is a useful website for exploring the science of climate modeling at a much more sophisticated level than I am assuming here. In particular, if you want to know why the effective layer thickness is proportional to the square root of the CO_2 concentration, look at scienceofdoom.com/roadmap/co2/ and read CO_2 parts three and four. Note that I am assuming that the "strong condition" applies (as the article claims is appropriate) and that the "optical depth" is proportional to the number of layers n. Part five discusses the 27% figure.

Where to get more information about the science of climate modeling

T10.7   Consequences

The problem with global warming is not so much the increase in temperature itself but rather the many and varied consequences of that increase. This is where things become even more uncertain (and more political). However, even here, scientific modeling and measurement can provide insight.

One of the already observed consequences is that extreme weather events are becoming more common. Figure T10.6 shows that the probability of extremely hot summer temperatures has risen dramatically in recent years, so that events that would have been extremely rare in the past are now commonplace. Wildfire frequencies seem to correlate with extreme temperatures, and therefore have gone up markedly in recent decades. For more about the increasing frequency of extreme weather events (including a color version of figure T10.6), go to www.giss.nasa.gov/research/briefs/hansen_17/.

Extreme weather events

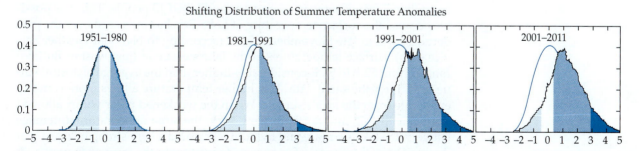

Figure T10.6

This figure shows how often summer temperatures are observed as a function of how much those temperatures deviate from the mean for the period 1951–1980. In the left graph, recorded temperatures are considered "cold" if they are in the left colored region, "typical" if they are in the center white region, and "hot" if they are in the right colored region, with a probability of 33% each (two sides of a six-sided die). By 2001–2011, the shift in the earth's temperature was sufficient so that "cold" temperatures only corresponded to one side of the die, "normal" to one side, and "hot" to four sides, with "extremely hot" (less than 0.5% probability before 1980) now amounting to nearly one half of one side of the die (8% probability, dark colored region on the right). (Credit: NASA/GISS)

Rising sea level

Another automatic consequence of a global temperature increase is an increase in the sea level, partly because water expands as it gets warmer, but mostly because of melting ice on land. The average sea level has already risen about 20 cm from the 1900 baseline, and is currently increasing at 3.3 mm/y. Scientists are casting worried eyes on the ice sheets in Greenland and in West Antarctica, which (if they were to collapse) could raise sea levels by perhaps 6 meters (20 feet), with severe consequences for many coastal cities. A paper published in 2014 in *Science* (**344,** 6185, pp. 735–738) presents some evidence that the collapse of the West Antarctic ice sheet has already begun (though it might take hundreds of years to complete).

More violent storms

If you have read chapter T9, then you know that major storms (such as hurricanes and typhoons) are heat engines that derive their power from temperature differences. Warming oceans therefore might plausibly yield more extreme temperature differences and so more extreme storms. There is evidence that storms of various types have become more powerful in recent decades. The 2013 report from the Intergovernmental Panel on Climate Change (IPCC) states (in section 2.6.3) that the evidence is "virtually certain" that power dissipation of tropical storms has increased in the northern Atlantic since the 1970s, and that a similar but less certain trend exists in the northern Pacific. While one cannot causally link this pattern to specific storms, 2012's Hurricane Sandy in the United States and 2013's Typhoon Haiyan (which had the strongest sustained winds ever measured) exemplify the kind of storms that the data and models suggest will happen more often in the future. A different study (Elsner et al., *Climate Dynamics*, DOI 10.1007/s00382-014-2277-3, 2014) documents a modest decrease in the number of tornados in the United States but a sharp increase in the *intensity* of tornado storm systems, particularly in the number of storm systems that create many tornados.

Cold weather events related to the so-called "polar vortex"

Ironically, global warming can lead to extreme local cooling. Published research (Kim et al., *Nature Communications* 5, 4646 doi:10.1038/ncomms5646) suggests that the weakening of the "polar vortex," which has been responsible for severe cold snaps in the United States during the winters of 2013 and 2014, arises from decreased insulating sea ice in the Arctic above Russia, which allows heat to flow from the sea to the atmosphere during early winter. The extra heat weakens the jet stream pattern (the polar vortex) that isolates cold Arctic air from middle latitudes, permitting oscillations in the jet stream that can carry very cold Arctic air into the United States and Europe. Days before I wrote these words in late November 2014, Buffalo, New York, experienced an unprecedented snowstorm (even for Buffalo) that laid down more than 7 feet of snow in some areas and killed 13 people. This was possible because the surface temperature of Lake Erie has increased over the past decades, and in late November of 2014 (according to National Weather Service data), surface temperatures at the lake's east end (just west of Buffalo) ranged from 52°F to 54°F, considerably higher than the average historical value of 47°F for the same date. This higher temperature allows more water to evaporate from the lake's surface. When the weakened polar vortex allowed Arctic air to spill into northern New York, the large temperature difference between the lake's surface and the cold air created a very strong blizzard that dumped record amounts of evaporated moisture in the form of snow.

Shifts in rainfall and drought

Indeed, the most worrying long-term consequences of climate change may be its effect on complex patterns of evaporation, rainfall, and drought. For reasons that you can understand from chapter T6, small increases in surface temperature on land or sea can *dramatically* affect evaporation rates (think about the Maxwell-Boltzmann distribution). This partly contributes to the intensity of storms (as noted above) but can also shift patterns of rainfall and drought. The IPCC report notes that one can have "medium confidence" that drought will increase in East Asia, and "high confidence" that drought will increase in western Africa and the Mediterranean, which will plausibly lead to increased political instability in these already unstable regions.

Moreover, important portions of the world's agriculture depends on rivers fed by mountain glaciers. Glaciers are in retreat worldwide as global temperatures increase. When the glaciers are gone and the rivers they feed become less reliable, people become vulnerable to agricultural disasters that can lead to famine, which can in turn lead to political instability and war. A particularly worrying illustration is the Indus River, which has its source in the glaciers of the Tibetan Himalayas, flows through northern India, and then into Pakistan. The water in this river is absolutely essential to agriculture in Pakistan, and yet India (Pakistan's historical enemy) controls the upper part of the river. As the glaciers disappear from the Tibetan highlands (which they are doing at a documented and rapid rate), India will almost certainly seek to divert more of the dwindling river to support its own agricultural needs. Perhaps you can imagine the geopolitical crisis that will likely result.

Failing agriculture can lead to political problems

The news during your lifetime will be filled with crises and trauma, wars and rumors of wars (as always), but if you probe deeply enough, you may find that the deep root of an increasing number of these crises is the simple, basic physics of climate change.

T10.8 Concluding Scientific Postscript

Understanding connections and foreseeing consequences such as these is what science is all about. I hope this chapter illustrates how physical model-building plays a crucial role in exposing the roots and connecting sinews of this very important real-life problem. Science *matters*. Without the basic physics that you have learned in this unit and this course, we would collectively be blind to this problem that seriously threatens our future.

Science matters

While I focused in the last section on climate change's hazardous consequences, you are also uniquely positioned by your knowledge of physics to contribute to the many and various ways that you can help the world avoid the worst consequences of climate change. Some possibilities include

Ways that you (with your physics knowledge) can help

- Developing new sources of non-carbon-based energy.
- Developing new kinds of batteries that make it easier to use intermittent renewable energy sources such as wind or solar power.
- Designing better computers and computer models for predicting effects.
- Creating and launching better earth-monitoring satellites.
- Finding ways to extract carbon from the atmosphere.
- Developing new ways to extract water from the air or ocean.
- Helping to educate people to understand climate change, and generally making it politically possible for our leaders to address the issues.

This is but a partial list of things that you *personally* can do to help, because you are among the less than 1% of the population that understands the science behind both the problem and many of its solutions. Science, with its unique balance of imagination, transnational community, skepticism, empiricism, and (above all) an absolute commitment to the truth, will play an absolutely essential role in any successful resolution of this problem.

The gifts you have bring responsibility. The gates to my college are engraved with the following: "They only are loyal to this college who departing bear their added riches in trust for [human]kind." Your gifts of understanding are a treasure that the world desperately needs. Our hope is that you will indeed bear your gifts to the world.

Bear your gifts in trust

TWO-MINUTE PROBLEMS

T10T.1 According to the argument in the text, the Moon's average surface temperature (since it has no atmosphere and is the same distance from the sun as the earth) should also be 255 K, or it would be if all points on its surface had the same temperature (and the Moon has the same average reflectance as the earth). But the Moon rotates very slowly relative to the sun (once a month) and has no atmosphere or ocean to redistribute energy differences. How do you think that these facts might affect the Moon's average temperature? (*Hint:* Suppose that half the moon was 100 K hotter and the other was 100 K colder than the average. Would it still radiate as much energy as it would if its surface area were all at the same temperature?)
A. The average should be less than 255 K.
B. The average should be more than 255 K.
C. The average should be the same: the energy flowing in must still balance the energy flowing out.

T10T.2 The earth's equilibrium temperature depends on its radius. T or F?

T10T.3 Doubling the number of effective opaque layers in a planet's atmosphere from n to $2n$ will make its average surface temperature increase by what factor?
A. 2
B. $2^{1/4}$
C. $3^{1/4}$
D. $(2 + 1/n)^{1/4}$
E. $[2 + 1/(n+1)]^{1/4}$
F. Some other factor (specify)
T. The factor depends on the planet's initial temperature.

T10T.4 If the atmosphere's "optical thickness" n depended on the ratio of the concentrations to the first power instead of to the 1/2 power in equation T10.8a, would this accelerate or brake the warming compared to the model in the text?
A. Accelerate
B. Brake
C. No change

T10T.5 According to the text's final model (including the ocean effect), about when will the earth's average surface temperature be 288.4 K?
A. 2015
B. 2030
C. 2045
D. Some other date (specify)

T10T.6 The model we have constructed in this chapter neglects feedback effects that might accelerate or brake the warming. Scientists now predict that the Arctic Ocean will be ice-free in the summer by as early as 2020 (due to increased temperatures in the Arctic). Less ice will mean that the Arctic will reflect less visible light back into space. Will this accelerate or brake the warming compared to the predictions of the model in the text?
A. Accelerate
B. Brake
C. No change

T10T.7 The model we have constructed in this chapter neglects feedback effects that might accelerate or brake the warming. It is possible that higher temperatures will yield increased evaporation and increased cloud cover. Will this accelerate or brake the warming compared to the predictions of the model in the text?
A. Accelerate
B. Brake
C. No change

T10T.8 Suppose scientists discover (somehow) that if ΔT, the average global temperature increase since 1900, were to exceed 1.0°C, Greenland's ice sheet would irreversibly start to collapse within 20 years, raising ocean levels significantly. ΔT is currently 0.8°C. What might we do to prevent this from happening?
A. Cut CO_2 emissions in half.
B. Completely stop emitting CO_2 now.
C. Actively start removing CO_2 from the atmosphere.
D. We probably can't prevent this from happening, no matter what we do.

HOMEWORK PROBLEMS

Basic Skills

T10B.1 Other things being equal, what would the earth's average temperature have been historically (before the Industrial Revolution) if it reflected only 25% of the light from the sun instead of 30%?

T10B.2 Other things being equal, what would the earth's average temperature have been historically (before the Industrial Revolution) if it reflected 35% of the light from the sun instead of 30%?

T10B.3 Some estimates suggest that if the carbon in all *currently known* reserves of fossil fuels were burned, the atmospheric concentration of CO_2 would peak above 540 ppm late in this century. If this is true, what would be the earth's average temperature at this peak concentration? (Assume thermal equilibrium.)

T10B.4 Suppose that we continue to add CO_2 to the atmosphere at a rate of 2 ppm per year until 2030 (which would correspond to *zero* growth in emissions), and then suddenly stop. What would our simple model predict that the earth's average temperature will be in 2060?

T10B.5 The scientific community's consensus is that an average temperature increase of more than 2 K above the pre-industrial average will cause irreversible damage to the earth's ecosystems. Suppose we continue to add CO_2 to the atmosphere at a rate of 2 ppm per year until we abruptly stop. According to the model described in this chapter, when must we stop?

T10B.6 If the lag time caused by the ocean were 15 years instead of 30 years (other things being equal), what would the earth's current (2015) average temperature be?

T10B.7 A planet with no atmosphere orbiting a sunlike star is tidally locked so that one side always faces the star. Suppose that side's uniform temperature is 355 K and the backside's is 155 K (an average of 255 K).
  (a) What average power per uni area does this planet thermally emit into space?
  (b) Compare to the 240 W/m^2 that the earth's upper atmosphere radiates at the same average temperature.

T10B.8 Using a computer, plot a graph of the earth's equilibrium average temperature as a function of CO_2 concentrations from 280 ppm to 800 ppm according to the model presented in this chapter.

Modeling

T10M.1 Consider a rapidly rotating planet without an atmosphere and a reflectance of about 12% (consistent with a rocky body like Mercury or the Moon). Construct a graph of the planet's average temperature as a function of distance from a sunlike star. (Express distances from the star in terms of AU, where 1 AU is the earth's average distance from the sun, and show the range 0.2 AU to 5 AU.)

T10M.2 How many effective opaque layers n of CO_2 would give Mars a pleasant average surface temperature of 14°C? (Mars is 1.52 times farther from the sun than the earth and reflects about 25% of the light incident on it.)

T10M.3 Predict the surface temperature of Jupiter's moon Callisto, and compare your result with the observed temperature, which you can find online. (Callisto has no atmosphere and reflects 20% of the light hitting it. Jupiter is 5.2 times as far from the sun as earth is.)

T10M.4 Suppose that a planet with no atmosphere orbits a sunlike star at the same radius as the earth's orbit and keeps the same face toward the sun at all times. The planet reflects 20% of the light falling on it, and the average temperature on the facing surface is nearly uniform and 200 K larger than the nearly uniform temperature on its back surface. What is the planet's *average* temperature? (*Hints:* Let the temperatures of the two sides be $T \pm \frac{1}{2}\Delta T$, where T is the average temperature and $\Delta T = 200$ K. You will have to solve what looks like an equation involving powers of T up to T^4, but if you look at it carefully, you will see that you can solve it using the quadratic formula.)

T10M.5 To understand why the ocean might have a big effect on slowing down the earth's temperature increase, consider the following. Scientists estimate that the current difference between the energy coming into the earth at present and the energy that would go out if the earth were in equilibrium is about 1.5 W/m^2. As calculated in the text, the current difference between the earth's actual temperature and the temperature we would expect in equilibrium is about 0.5°C. The ocean has an average depth of about 3700 m, a density of about 1025 kg/m^3, and a specific heat of about 3850 J/(kg·K). If each square meter of ocean surface absorbs energy at a rate of 1.5 W/m^2, about how long would it take for the entire water below to warm by 0.5°C? (*Hint:* The ocean's area does not matter—why? Don't be alarmed if your result is larger than 30 years: one does not need to warm the *whole* ocean *uniformly* from top to bottom to cause the earth's average surface temperature to increase by 0.5°C.)

T10M.6 One possible solution to the climate change problem would be to increase the earth's average reflectance (the technical term is **albedo**) above 30%.
  (a) What albedo would we need to keep the earth at its historical average temperature of 14°C but with a peak of 550 ppm of CO_2 in the atmosphere?
  (b) Roughly what fraction of the earth's area would we have to paint bright white to make this happen?
  (c) Express this area in square kilometers, and compare it to the area of the United States.

T10M.7 Suppose that in 2016 we start reducing the net CO_2 we add to the atmosphere from the 2015 value of 2 ppm/y by 0.04 ppm/y every year, that is to 1.96 ppm in 2016, 1.92 ppm in 2017 and so on (this, initially, would mean reducing total carbon emissions by about 2% per year).
  (a) Show that this means we will stop adding net CO_2 to the atmosphere entirely in 2065.
  (b) What will be the concentration of CO_2 in 2065? (*Hint:* Draw a graph, and use the graph to calculate an integral which otherwise might be modestly difficult.)
  (c) According to the simple model in this chapter, what will be the earth's peak average temperature?
  (d) Is this under the limit described in exercise T10X.3, or do we need to be more aggressive?
  (e) When will the earth's average temperature reach its peak value?
  (f) What will happen thereafter?

T10M.8 Water molecules evaporate from a body of water when random collisions between molecules happen to give a few molecules kinetic energies large enough exceed the binding energy that holds water molecules together, which is about 0.42 eV. As a rough approximation, model the water molecules as gas molecules obeying the Maxwell-Boltzmann distribution. Use the MBoltz app to estimate how many times more water molecules have enough energy to evaporate at $T = 15$°C than have enough energy at $T = 14$°C.

Rich-Context

T10R.1 Suppose that astronauts discover a rapidly spinning metal cube with wavelength-independent emissivity $\epsilon = 0.3$ orbiting the sun at the same distance as the earth is from the sun. Assume that the cube's rotation axis is perpendicular to the plane of its orbit. What is the cube's average surface temperature? (*Hint:* You need to find the cube's average cross-sectional area, which will involve doing a simple trigonometric integral. Also note that metals are good conductors of heat.)

T10R.2 One solution to the crisis of climate change would be to increase the earth's orbital radius. Suppose that as we finish with burning fossil fuels, the concentration of CO_2 stabilizes at about 550 ppm. We would like to maintain the earth's average temperature at 14°C.
 (a) By about how much would we have to increase the earth's orbital radius to achieve this goal?
 (b) How much energy would be required to do this?
 (c) Compare this to the total solar energy that the earth receives in a year.

T10R.3 Suppose that in 2015 we start reducing the net carbon we add to the atmosphere linearly from the current value of 2 ppm/y to zero at some future date. According to the model developed in this chapter, what does that date need to be to avoid exceeding the 289 K limit that most scientists think is when damage to the ecosystem will be extensive and irreversible? (*Hint:* See problem T10M.7 to help you develop an outline for how you might proceed.)

T10R.4 In Larry Niven's classic science fiction novel *Ringworld* (1970), alien engineers construct a huge ring like a thin band that encircles a sunlike star. The ring has a surface area of several million earths that faces the star, and an equal surface area that faces away. The ring rotates at such a rate so that the (fictitious) centrifugal force in the frame of its surface is equivalent to earth's gravity. High walls at the ring's edges confine an atmosphere like earth's, with 280 ppm CO_2. Assume that the ring's inner surface, like the earth's, reflects 30% of the star's light incident on it.
 (a) Assume that the ring is thin enough that the temperature on its backside is essentially the same as that on its front surface (the surface facing the star). What should its diameter be (as a multiple of the diameter of the earth's orbit) if the average temperature on the ring's front surface is 14°C?
 (b) Assume that the ring's backside is instead 100 K cooler in equilibrium than its front surface. What should the ring's diameter be now?

ANSWERS TO EXERCISES

T10X.1 With $n = 256$, we have $(n + 1)^{1/4}T_0 = (257)^{1/4}(184 \text{ K})$ $= 737 \text{ K}$, as claimed.

T10X.2 According to our model, the surface temperature in 2050 will be the same as the equilibrium temperature for the 2020 CO_2 concentration, which will be the 2015 concentration of 400 ppm + (5 y)(2 ppm/y) = 410 ppm. According to the methods in section T10.4, we will have

$$n = 0.163\left(\frac{410}{280}\right)^{1/2} + 0.442 = 0.639 \qquad \text{(T10.11}a\text{)}$$

$$T = (n + 1)^{1/4}T_0 = (1.639)^{1/4}(255 \text{ K}) = 288.5 \text{ K} \quad \text{(T10.11}b\text{)}$$

or about 1.5°C above the baseline temperature.

T10X.3 According to our model, the peak allowable value of $n + 1$ is such that $T = (n + 1)^{1/4}T_0 = 289 \text{ K}$. This means that we must have

$$n = \left(\frac{T}{T_0}\right)^4 - 1 = \left(\frac{289 \text{ K}}{255 \text{ K}}\right)^4 - 1 = 0.650 \qquad \text{(T10.12)}$$

But this requires that the maximum CO_2 concentration (call it C) must be such that

$$n = 0.163\left(\frac{C}{C_0}\right)^{1/2} + 0.442 \qquad \text{(T10.13)}$$

where $C_0 = 280$ ppm. Solving for C yields

$$\left(\frac{C}{C_0}\right)^{1/2} = \frac{n - 0.442}{0.163} = \frac{0.650 - 0.442}{0.163} = 1.276$$

$$\Rightarrow C = (280 \text{ ppm})(1.276)^2 = 456 \text{ ppm} \qquad \text{(T10.14)}$$

So, according to our crude model, this is the *most* CO_2 we can allow the atmosphere to have. At the current rate of 2 ppm/y, we will reach this concentration in 2043 (and the earth temperature will hit 289 K in 2073).

Index

NOTE: Page numbers followed by *f* and *t* indicate figures and tables, respectively.

Periodic Table of the Elements

1 1A																	18 8A
1 **H** 1.008	2 2A											13 3A	14 4A	15 5A	16 6A	17 7A	**2** **He** 4.003
3 **Li** 6.941	**4** **Be** 9.012											**5** **B** 10.81	**6** **C** 12.01	**7** **N** 14.01	**8** **O** 16.00	**9** **F** 19.00	**10** **Ne** 20.18
11 **Na** 22.99	**12** **Mg** 24.31	3 3B	4 4B	5 5B	6 6B	7 7B	8	9 8B	10	11 1B	12 2B	**13** **Al** 26.98	**14** **Si** 28.09	**15** **P** 30.97	**16** **S** 32.07	**17** **Cl** 35.45	**18** **Ar** 39.95
19 **K** 39.10	**20** **Ca** 40.08	**21** **Sc** 44.96	**22** **Ti** 47.88	**23** **V** 50.94	**24** **Cr** 52.00	**25** **Mn** 54.94	**26** **Fe** 55.85	**27** **Co** 58.93	**28** **Ni** 58.69	**29** **Cu** 63.55	**30** **Zn** 65.39	**31** **Ga** 69.72	**32** **Ge** 72.59	**33** **As** 74.92	**34** **Se** 78.96	**35** **Br** 79.90	**36** **Kr** 83.80
37 **Rb** 85.47	**38** **Sr** 87.62	**39** **Y** 88.91	**40** **Zr** 91.22	**41** **Nb** 92.91	**42** **Mo** 95.94	**43** **Tc** (98)	**44** **Ru** 101.1	**45** **Rh** 102.9	**46** **Pd** 106.4	**47** **Ag** 107.9	**48** **Cd** 112.4	**49** **In** 114.8	**50** **Sn** 118.7	**51** **Sb** 121.8	**52** **Te** 127.6	**53** **I** 126.9	**54** **Xe** 131.3
55 **Cs** 132.9	**56** **Ba** 137.3	**57** **La** 138.9	**72** **Hf** 178.5	**73** **Ta** 180.9	**74** **W** 183.9	**75** **Re** 186.2	**76** **Os** 190.2	**77** **Ir** 192.2	**78** **Pt** 195.1	**79** **Au** 197.0	**80** **Hg** 200.6	**81** **Tl** 204.4	**82** **Pb** 207.2	**83** **Bi** 209.0	**84** **Po** (210)	**85** **At** (210)	**86** **Rn** (222)
87 **Fr** (223)	**88** **Ra** (226)	**89** **Ac** (227)	**104** **Rf** (257)	**105** **Db** (260)	**106** **Sg** (263)	**107** **Bh** (262)	**108** **Hs** (265)	**109** **Mt** (266)	110	111	112	(113)	114	(115)	116	(117)	

58 **Ce** 140.1	**59** **Pr** 140.9	**60** **Nd** 144.2	**61** **Pm** (147)	**62** **Sm** 150.4	**63** **Eu** 152.0	**64** **Gd** 157.3	**65** **Tb** 158.9	**66** **Dy** 162.5	**67** **Ho** 164.9	**68** **Er** 167.3	**69** **Tm** 168.9	**70** **Yb** 173.0	**71** **Lu** 175.0
90 **Th** 232.0	**91** **Pa** (231)	**92** **U** 238.0	**93** **Np** (237)	**94** **Pu** (242)	**95** **Am** (243)	**96** **Cm** (247)	**97** **Bk** (247)	**98** **Cf** (249)	**99** **Es** (254)	**100** **Fm** (253)	**101** **Md** (256)	**102** **No** (254)	**103** **Lr** (257)

Atomic number

1 **H** 1.008 — Symbol

Atomic mass

Short Answers to Selected Problems

[Note that most of the derivation (D) problems, as well as a number of other problems, have answers given in the problem statement. These problems are also useful for practice, but their answers are not reiterated here.]

Chapter T1
B1 (a) R, (b) I, (c) I, (d) R, (e) I. **B3** (a) Q, (b) $[E]$, (c) W, (d) W, (e) $[E]$. **B5** 0.266 kg. **B7** (b) 531.7°R. **B9** −56.7°C, 216.5 K. **B11** 125 J. **M3** 361 K. **M5b** 131 W. **M7** 9.7°C. **M9** 220 J·kg^{-1}K^{-1}. **D4a** $3k_B N_A/M_A$. **R1a** The first thermoscope is bad.

Chapter T2
B1 28. **B3** 352,716. **B5** 3.6. **B7** 0.2×10^{-6}. **B9** 5.4×10^{42}. **M1b** $U_A = 1$, $U_B = 5$ and $U_A = 2$, $U_B = 4$; yes in the latter case. **M3b** $U_A = 4$, $U_B = 5$ and $U_A = 5$, $U_B = 4$; close in both cases. **M5b** $U_A/N_A \approx U_B/N_B$. **M7b** 10,000 km. **M9** 331 times. **R1b** $\Omega(U, N) = N!/(U/\varepsilon)!(N - U/\varepsilon)!$.

Chapter T3
B1 $646k_B$. **B3** (a) $2378.5k_B$, $2501.3k_B$ (b) $4880k_B$. **B5** $S > 69$ J/K. **B7** 350 K. **B9b** 352 K. **M1** Cleaning still generates entropy. **M3** 2.0×10^{13}. **M6b** Atypical but "normal." **M7** $U = \frac{3}{2}Nk_B T$. **M9** (a) 1.22 J/K, (b) 737 J/K. **R1c** $1/T = k_B[N - 2(U/\varepsilon)]/U$.

Chapter T4
B1 (a) 290 K (b) 1.7%. **B3** 4.0×10^{-6}. **B5** 602, 274, 124. **B7** 1.157. **B9a** 0.12 eV. **B11** ε. **M1** 0.99952, 0.00048. **M3** 1.163. **M5** Cool it to 200 K. **M7** 1.67 K. **M9** 0.9973. **D6** $3Nk_B(\varepsilon/k_B T)^2 e^{-\varepsilon/k_B T}$. **R1** (a) 2.4 K (b) likely, given uncertainties. **R3** $U = 2N\varepsilon/(e^{2\varepsilon/k_B T} - 1)$.

Chapter T5
B1 Z increases. **B3** 1.63×10^{-20}. **B5** (a) 0.8 mK, (b) 1.2 K, (c) 1100 K. **B7** (a) 370 J, (b) 3.2 m/s. **B9** (a) about 1.5×10^{25}, (b) No. **B11** 7700 K. **B13** 92 K. **M1c** 0.17. **M3** mostly off. **M5b** 0.62 K. **M7** 100,000 Pa. **D7b** $\frac{3}{2}k_B T$. **D9c** 0.33. **R1c** diameter $\approx$ 15 m. **R2e** $\approx$ 2.4 h. **R3c** 0.33 times the pressure at sea level.

Chapter T6
B1 2.38×10^{-23} kg·m/s. **B3** 1.72×10^{22}. **B5b** 0.0059. **B7** 0.572. **B9a** 5.7 µJ/m^3. **B11** 6.0×10^{-5}. **B13** 0.00955. **B15** 6. **M1** 2.46×10^{14} ions/s. **M3** 0.00226. **M4d** 980 Cal/d. **M7** Photon gas will be exactly the same. **R2b** about a factor of 2. **R3b** 2.32×10^{16}/s.

Chapter T7
B1 (a) 3.01×10^{23}, (b) 3060 J, (c) 61,300 Pa. **B3** +0.29 J. **B5** Heat must flow out of the gas. **B7** +69 J. **B9** −24 J. **B11** 97 kPa. **M1b** 100 J. **M3** 130 J. **M5** first row: 0 − −, second row: − + 0, third row: + 0 +. **M7** 55 cm^3, −68°C. **M9** (a) monatomic, (b) 374 J. **M11** diatomic. **D4** $TP^{(1-\gamma)/\gamma}$ = constant. **R1** About −5°C. **R3** 28.3 kJ.

Chapter T8
B1 $S = Nk_B \ln[(V/N)(2\pi e^{5/3} \; mk_B T/h^2)^{3/2}]$. **B3b** +0.04 J/K. **B5** +5.4 J/K. **B7** +760 J/K. **M1** 10^{-x} where $x = 3.01 \times 10^{22}$. **M3** (a) −215 J/K, (b) +243 J/K. **M5** +3.7 J/K. **M7** 10^{-x} where $x = 2.2 \times 10^{20}$. **M9** (a) no, (b) no, (c) yes. **D4b** $\Delta S = 2Nk_B \ln 2$, where N is the number of molecules in each half. **D5a** $15.1Nk_B$. **R1** roughly 80 J/K. **R3** very roughly 8000 J/K.

Chapter T9
B1 yes. **B3** 0.27. **B5** It needs a place to discard heat. **B7** 17.2. **M1** A hydroelectric plant is not a heat engine. **M5** (a) $|Q_H|/W$ (b) $T_H/(T_H - T_C)$. **M9** 860 kg. **M11** 135 kg/s. **M13** (a) 258 J, (b) uses only 37% of the heat required to warm the soup directly. **D2** Decrease the temperature of the cold reservoir. **D5d** 0.56. **R1** The Minnesotan saves 1.7 times more than the Californian. **A1** 2.2 liters, 0.22 liters.

Chapter T10
B1 4.7 K hotter. **B3** 289.8 K. **B5** 2043. **B7a** 467 W/m^2. **M2** $n \approx 2.47$. **M3** 116 K. **M5** 154 y. **M7** (b) 450 ppm, (c) 289 K, (e) 2095. **R1** 267.4 K. **R3** 2055.